The City in the Experience Econo

The book develops a new approach to urban development in which leisure, pleasure or experiences are seen as key drivers. History, authenticity, urban qualities, local culture and leisure offerings or a vibrant retail sector are thus assets in local development also outside of the big cities. Globalization and high mobility are necessary aspects of the development, which entails the development of high urban profiles in a globalized and highly competitive world. Apart from experiential qualities a critical urban size is also required. Experience qualities can be connected to urban design, where particular designs stimulate citizens' learning and activity in the urban space. They can also be connected to more tourist related large scale projects of experiential mass consumption with fun parks and shopping. A combination of the two approaches has been developed to promote, for example, car brands and cities through experiential car museums. New stakeholders, new network based forms of cooperation and new entrepreneurial strategies are connected to urban development in 'the experience economy'. In particular new network based approaches are needed if small and rural places should also reap the fruits of the experience economy.

This book was originally published as a special issue of *European Planning Studies.*

Anne Lorentzen is Professor in Geography at the Department of Development and Planning, Aalborg University, Denmark. She has a PhD degree in Technology Planning.

Carsten Jahn Hansen is Associate Professor in Planning at the Department of Development and Planning, Aalborg University, Denmark. He has a PhD degree in Urban Planning and Urban Development.

The City in the Experience Economy
Role and Transformation

Edited by
Anne Lorentzen and Carsten Jahn Hansen

Routledge
Taylor & Francis Group

LONDON AND NEW YORK

First published 2012
by Routledge
2 Park Square, Milton Park, Abingdon, Oxon, OX14 4RN

Simultaneously published in the USA and Canada
by Routledge
711 Third Avenue, New York, NY 10017

First issued in paperback 2014

Routledge is an imprint of the Taylor & Francis Group, an informa business

British Library Cataloguing in Publication Data
A catalogue record for this book is available from the British Library

ISBN13: 978-0-415-69734-7 (hbk)

ISBN 13: 978-1-138-85345-4 (pbk)

Typeset in Times New Roman
by Taylor & Francis Books

Publisher's Note
The publisher would like to make readers aware that the chapters in this book may be referred to as articles as they are identical to the articles published in the special issue. The publisher accepts responsibility for any inconsistencies that may have arisen in the course of preparing this volume for print.

Contents

INTRODUCTION

The Role and Transformation of the City in the Experience Economy: Identifying and Exploring Research Challenges

ANNE LORENTZEN & CARSTEN JAHN HANSEN

ABSTRACT *The article introduces the special issue on the role and transformation of the city in the experience economy and raises some research issues. The issue focuses on the transformative aspects that can be identified in relation to experience oriented planning and development. We show how experience economy is much related with affluence and the growth of leisure demand, and that place has a particular role to play in terms of amenities, narratives and identities. Places, and in particular cities, undergo development or commodification to attract leisure consumers, which are increasingly mobile. Both urban systems (the relative position and role of cities) and urban structures (the city fabric) change in the experience economy, and so does means and ends of planning, which can be seen to be increasingly entrepreneurial and stakeholder based.*

Introduction

This special issue of *European Planning Studies* explores and discusses the emerging experience economy and its implications to economic geography, physical spaces, mobility, environment, culture, branding, planning and democracy. In particular, our research focuses and conceptualizes on the transformative aspects that can be identified in relation to experience-oriented urban development. Our overall hypothesis is that a range of different places are becoming integrated in the experience economy through the development of still more challenging leisure activities, new advertising and marketing strategies and through the capitalization and development of places. Rural and remote areas are, to some extent, part of the trend as tourist resorts or as locations for the production of specialties for the global market. However, cities are mostly involved due to the greater variety and density of activities and places hosted by them. This is illustrated by the fact that an increasing number of cities have embarked on strategies of experience-oriented growth. Hence, this issue focuses on the role and transformation of the city in the experience economy.

Our exploration departures from and involves different interrelated spatial perspectives. In the outset, we ask what implications the development of experience-based activities has for scales of economic geography. Does the emerging experience economy represent new opportunities for cities of different sizes and positions in the urban hierarchy? Furthermore, among the broad range of what has been termed "experience industries" (Tofler, 1970), we particularly consider place-bound or attendance-based industries because of their immediate role for local development, as well as for urban policy and planning. Hence, since planning is a force which is decisively intertwined with geographies, we have also asked: what is experience planning? Through what kind of strategies, projects and activities do planners contribute to the experience content of their local territory? and vice versa; how does experience-oriented development influence local policy and planning activities, and what are the physical, cultural and democratic consequences of this? In doing so, we consider and discuss several more specific consequential aspects, such as the relationship between the experience economy, the mobility of experience customers and the environment. Is there perhaps a contradiction between the low material intensity of experiences and the polluting global transportation? In addition, we explore into the characteristics and activities of stakeholders of the experience economy and the governance settings and practices related to experience-oriented planning. What are, for instance, the roles of civic, commercial and public sector stakeholders; do we see the formation of new relations and arrangements as well as a new topography of cooperation or "rules-of-the-game"?

The Growth of Leisure Economies

In his book *"Future Shock"*, Toffler (1970) foresaw radical changes in the economic structures of the advanced economies. Under more affluent conditions, Tofler (1970, p. 220) suggests, we are recognizing the economy to deal with a new level of human needs. According to Tofler, the economy will become geared to the provision of "psychic gratification" and the "quality of life". These changes are multifaceted, but since the basis of them is increasing affluence, it is relevant and useful to consider the history of this affluence, because it implies that today leisure and entertainment represent an important and increasing part of consumption patterns. Recreational goods and services represent luxury goods, which in economic terms is defined as goods with "income elasticity greater than unity". By rising incomes, the share of such goods in consumption will rise. And, incomes have been rising in the industrialized countries for a long period of time. Between 1870 and 2002, this growth was 2.3% annually as a mean. In addition, the share of recreational consumption grew considerably as part of disposable incomes. An estimate of the OECD concludes that between 1975 and 2002 it grew from 8.1% to 10.3% in Denmark and in the UK from 5.9% to 13.2% (Andersson & Andersson, 2006, p. 42).

However, not only rising incomes, but also an increase in leisure time contributes to the growth in recreational consumption. It has been shown that, in the OECD countries, the number of annual working hours has been almost halved between 1870 and 1979 (Andersson & Andersson, 2006, p. 45). The shortening of the working day and the increase in the average vacation time contribute to this. This means that the constraints to leisure consumption related to income and time have been considerably relaxed. Never before have the opportunities for the consumption of leisure products and services been so great, just as expected by Tofler. Furthermore, demographic and educational changes contribute to the growth in leisure demand. In the advanced countries, people live still longer, and the average age of the population is likely to increase in the future. In addition, the birth rate drops, leaving more money in each household for leisure consumption (Andersson & Andersson, 2006, p. 48).

Then, what do people demand, and in particular, what influences this demand? The structure of demand for leisure activities depend on people's education. This is because some leisure products and services require certain knowledge to be enjoyed. Generally, in OECD countries, the education level has increased considerably, from 2–3 years of school in 1870 to 11.6 years of school in 2000. This would, for instance, indicate a higher demand for knowledge-requiring consumption such as arts. Education levels still vary, and it has been shown that highly educated people tend to prefer museums, cinemas, live music performances and literature among others, whereas people with little education prefer amusement products and video recordings among others (Andersson & Andersson, 2006, p. 49).

Hence, the combined impact of income, time, demography and education factors means that leisure products and services represent a very important growth sector of the economy.

How Do We Conceptualize the New Development?

In economic terms, leisure products and services are then luxury products that people consume when they have a surplus of income and time, and (as indicated) the type of products preferred depends on the age, educational profile and family situation of the consumer. Tofler (1970, p. 227) provided "experiential production" as a term for the new segment of production directed towards luxury consumption. The industries were "experience industries" (Tofler, 1970, p. 221) and the producers "experience makers" (p. 219). Future consumers would be needing products and services that add beauty, prestige, individualization and sensory delight to the products or services (Tofler, 1970, p. 224). Experiences will even be sold on the market exactly as if they were things (Tofler, 1970, p. 226). "Psychic gratification" is the core of the new provision system (Tofler, 1970, p. 220).

In the 1990s, the sociologist Schulze (2005, p. 13) labelled the increasing focus on the non-material aspects of consumption as the aesthetication of everyday life. The significant role of the aesthetic appearance of products, human relationships and human habits shows how a larger part of people's lives has turned into experience projects and the society into an experience society. Therefore, Tofler interprets the new trend as a new provision system, whereas Schulze (2005) sees it as a matter of lifestyle, or rather as related to the meaning of life. Recently, these ideas have also been developed by Lund *et al.* (2005) and by Vetner and Jantzen (2007). Experiences become important to the individual because they become part of the narrative connected to his/her life and identity. In the age of modernity, it is part of the freedom of the individual to create his/her own identity, and experiences become a means to achieve this.

It was, however, the contribution of Pine and Gilmore (1998, 1999) that caused the rouse of a broader discussion and agenda-setting concerning the role of experiences. They seem to have been inspired by Tofler's original ideas and developed them into a strategic tool of business development. In addition, they apply Porter's idea that producers who apply more refined strategies of specialization, instead of just competing on price, will gain a more sustainable advantage on the market (Porter, 1990). Pine and Gilmore (1998, 1999) find that providers of goods and services increasingly compete on the basis of particular experience dimensions added to their product or service. The involvement of the consumer is a quality of the experience product or service: "ing the thing" e.g. canoeing, seal-viewing and dog-sledding (Pine & Gilmore, 1999, p. 15). From Tofler (1970, p. 225), they borrow the idea of memorable experiences. Success of a product or service is when the consumer keeps a pleasant memory of his/her meeting with the product or service. Two business perspectives can be deducted from their contribution. First, staging experiences becomes a business in itself in the future (Pine & Gilmore, 1999, p. 67), as exemplified by the increasing supply of leisure activities such as theme parks. And secondly, as a strategic management device experiences will become part of marketing or a potential add-on to all goods and services for final consumption, as seen in the efforts to aestheticize products and develop interesting relating narratives. In parallel to this, place branding can be understood as a strategy involving the aesthetication of places, the development of narratives about them and as the staging of the city as an experience.

What is an Experience Product?

Based on the above, experience products can be considered luxury products, which are consumed for pleasure. They are products that represent a narrative and some degree of involvement of the consumer, as an individual or as a group. Physical products as well as services can be enacted as experience products. These are either foot-loose (computer games), place-bound or attendance-based (a theatre play, a meal at a theme restaurant). As mentioned, this issue focuses on place-bound experiences, as places and activities can also be capitalized as experience products, directly or indirectly. Places may host experience services (stadiums, hotels) or they may in themselves represent attractive narratives, which motivate people to visit them (a historic castle, street of city).

However, common to the different categories of experience products is the appeal to the feelings of the consumer (Schulze, 2005, p. 421; Tofler, 1970). The definition of experience products is concerned with the relation between the consumer and the product. The litmus test is then if it appeals to the customer by creating an experience? This is not an immediately operational definition, and when empirically researching into issues of experience economy, researchers need to develop more specific approaches, depending on the context. What people, or rather consumers, consider an experience (a positive enjoyment (Jantzen, 2007)) is bound to be culturally, historically as well as sociologically specific. Therefore, it makes sense to focus on particular places, in particular historical periods when digging into the spatial importance of the experience economy. The empirical backbone of this issue is thus recent developments in Europe, and in particular, in Denmark.

What Does this Mean to Spatial Development?

It is not difficult to deduce the fact that leisure demand, luxury demand and the search for experiences is bound to be mostly concentrated in cities. One reason is that, in general, incomes are higher in or near the largest cities. Additionally, in cities, education levels are generally higher and families are generally smaller, often due to more career-oriented lifestyles. Furthermore, in a historical urban development context, experience- or leisure-oriented activities tend to have been concentrated in cities. Urban dwellers, therefore, should tend to spend more money on leisure products and services than people in rural areas, and they should tend to purchase more knowledge-demanding types of leisure products, than rural dwellers would. It can be added that many experience productions represent considerable economies of scale. They may require large investment in physical amenities (concert halls, theatres, sport stadiums) and they require a very large and stable public to create sufficient income. Also, the creativity involved in developing experience products benefits from what we call economies of urbanization (Andersson & Andersson, 2006, p. 183). This refers to the diversity and the density of cities, which implies that the likelihood of creative entrepreneurs to spontaneously encounter tacit knowledge and learning is high (Desrochers, 2001). It has been shown how many of the worlds creative neighbourhoods exhibit particular conditions of diversity and density (Andersson & Andersson, 2006; Jacobs, 1985; see also Florida 2005).

However, experience economic growth and innovation may also take place beyond the metropolises (Bell & Jayne, 2006), based on local initiative and public–private partnerships. One example is the successful branding of the Swedish region of Skåne as a food region, with the logo of "the spirit of food" (Baltic Sea Solutions, 2006). This logo was based on an emerging food cluster, the development of a food university and the active development of related narratives. In the INTERREG programme, research has been undertaken to analyse and support the role of small and medium-sized historic towns in Europe. The conclusion is that these towns bear a particular potential for stimulating regional competition through supplementing culture and leisure functions, which can serve urban development in economic as well as non-economic terms. The towns provide soft location factors for investors, they are attractive for tourism in the experience economy and, finally, they are anchor points for local and regional identity. Such towns may even serve as magnets for the creative and talented people who are the basis for creative industries. This is seen as a perspective even for peripheral, rural areas. One point, however, which is of utmost importance to such towns is their physical connectivity, but perhaps even more so their relational connectivity (Nagy *et al.*, 2007). Being physically connected to other places through high-rated transport infrastructures, such as motorways and airports, does not guarantee success. Additional relations are often required for a town or small city to be able to compete and develop.

This is also why another important aspect is the positioning of the city on the "market". The city is only attractive if it is not like any other city. It has to have a different profile. The positioning of cities is not as easy as the positioning of products, due to the many stakeholders of the city. The positioning of the city through branding is, therefore, both an internal process through which key stakeholders need to agree on the identity or image they want to promote, as well as an external process of communication with the different "markets" (Christgau & Jacobsen, 2004; Løkke, 2006). Similar to this discussion, Zukin (1995) suggests that cities have a "symbolic economy" on the basis of which they compete. The symbolic economy is the look and feel of cities. Since the 1970s and following the industrial decline, the image of the city has become a product that is sold on the national and even the global level. Private enterprise is responsible for a large share of the offer, of which much is "entertainment", aiming at attracting a mobile public of cultural consumers (Zukin, 1995, p. 19). This changes the culture, the cultural policy and the urban development policy of cities. Instead of being a welfare good, culture becomes a globally marketed good (Freestone & Gibson, 2006). See also the next section.

Furthermore, Romein (2005) discusses the relationship between, on the one hand, the development of the urban system and, on the other hand, the activity system of leisure. An increasing part of the urban economy consists of the production of symbolic goods, such as entertainment and decoration, just as Tofler, Schulze and Pine and Gilmore claimed concerning the economy in general. Much of this is attendance-based, as it needs to be consumed *in situ*. By producing such products, the cultural (or experience) industries potentially contribute to both the quality of leisure and the quality of place in cities. This can be to the benefit of the citizens of the particular place. However, there is also a risk that it may contribute to, for instance, a theme orientation in urban spatial development that excludes certain citizen groups or creates new social divides between citizens.

Another aspect of importance to urban spatial development, or the urban system, is that today people invest time and money in travel to pursue their leisure interests. The attendance-based and place-bound characteristics of many experiences mean that leisure consumption and travel are often closely connected. This influences mobility patterns, often through increases in mobility, which again puts pressure on urban infrastructures and the environment. In addition, the combination of increasing mobility and the individualization of tastes implies the emergence of new communities of shared preferences regarding consumption and leisure. This again may imply the emergence of a certain division of labour between leisure activity suppliers in different places. Certain outdoor activities, such as wind surfing and skiing, concentrate in particular places depending on the natural circumstances. Some cities are known for music festivals of rock (Roskilde) or classical music (Århus, Salzburg), shopping (London) or amusement parks (Billund).

The consequences of the above reflections on urban spatial development are that the centre of large cities tend to transform into spaces of consumption for fun and enjoyment (Mullins *et al.*, 1999). However, the establishment of shopping malls, mega dance halls, sport stadiums, multiplex cinemas, integrated entertainment complexes and amusement parks also takes place outside the inner cities. They can be seen at city edges, in suburbs, along highways and in well accessible small towns or villages. The variety of places for entertainment and leisure is indeed quite impressive today. There is no longer a particular urban centre to go to. Romein (2005, p. 15) suggests the idea of a polycentric "leisure field", offering a large diversity of supplementary amenities and experiences. The large and diversified offer has two main implications. First, due to high mobility, the space of leisure consumption of the individual family is extensive. Often, consumers do not depend on the proximity of the specialized offer of leisure activities, because they are willing to travel to be *in situ*. Residential location is not necessarily linked very closely to the spatial distribution of leisure consumption spaces. Second, people also increasingly tend to demand the quality of places (Healey, 2007), e.g. of a vibrant city with a large diversity of amenities and experiences. In a developmental perspective, this gives large cities a competitive edge over small cities and peripheral places, due to the mentioned economies of scope and scale of the offer of leisure activities.

However, it also means that small cities may achieve specialized roles and competitive advantages in a polycentric network of outdoor leisure supply, because people are willing to travel to have experiences. Furthermore, it means that cities of all sizes need to focus on the quality of place to keep up or expand the number of citizens as well as visitors or tourists (Bradley & Hall, 2006; Bell & Jayne, 2006). The findings of Markusen and Schrock (2006) show how urban resurgence or growth is connected to the distinctiveness of cities and not necessarily to their size. It can be shown how the success of cities is related to their distinctive production structure, their consumption mix and their identity/citizenship, or in other words to their cultural uniqueness (Markusen & Schrock, 2006, p. 1302). This also means that even smaller cities compete internationally (Markusen & Schrock, 2006, p. 1306). However, big cities at the top of the urban hierarchy still show traditional strength in relation to the arts, media and architecture (Markusen & Schrock, 2006, p. 1318). Research thus suggests that there are important spatial dimensions related to the experience economy. Based on this, we deduce that city size and structure, urban qualities and urban systems are all in play in the transformation of the city in the experience economy.

What Does this Mean to the Governance and Planning of Cities?

When focusing on the quality of places and identity and cultural uniqueness as competitive parameters, we also ask: how do cities handle that; what urban governance settings and practices can be identified in relation to the experience economy? Changes in prevailing urban development conditions are likely to be met or matched by a changing governance system. Governance systems may change or respond to a variety of pressures, such as globalization, a concern for environmental qualities, shifts in lifestyle and cultural values (Healey, 2007). Historically, it has often been economic imperatives that set the stage for urban development and governance. As the orientation towards leisure and experiences may imply added economic value to urban spaces, the specific contents of strategies and plans for urban development can also be expected to change alongside changes in governance.

Healey (2007) suggests that the general processes of urban strategy-making and planning are changing. In itself, a shift in development focus, themes and specific projects brings along new actors, stakeholders and arenas for policy formation, which is likely to influence governance processes (Hajer & Wagenaar, 2003). In addition, more attention is given to provide appropriate spaces for new kinds of production, commercial, financial and consumption activity. This leads to a transformation dynamic that seeks to swing established governance processes "locked in" to old integrated and "managerial" modes of governance towards more "entrepreneurial" approaches to developing the assets of urban areas (Healey, 2007, p. 23). Today, communities and companies all over the world are replacing hierarchies with networks, authority with empowerment, order with flexibility and creativity and paternalism with self-responsibility (Landry, 2006, p. 292). The idea seems to be to generate and share more resources (economic, knowledge, institutional, etc.) as well as to achieve more effective implementation. Planning this has implied a move away from its land-use focus towards being more about mediation and the negotiation of differences (Landry, 2006, p. 299). The implication is that urban areas cannot be "planned" by government action in a linear way, from intention to plan, to action, to outcome as planned. What goes on in urban areas is just too dynamic, intricate and mazy (Healey, 2007, p. 3). This has entailed a search for new settings and practices that embrace, link and facilitate combined or joined-up efforts among a more complex and varied range of actors and interests. This again implies a move from bureaucratic welfarist organization and distribution towards local growth policies as well as partnering and more entrepreneurial approaches, in which even traditional welfare and land-use values are becoming integrated in competitive place development and branding. The development of the quality of urban place in a competitive perspective does, however, run the risk of social exclusion and fragmentation, since not all urban dwellers fulfil the income, time and education requirements to participate in the experience economy.

This Issue

Our research into the role and transformation of the city in the experience economy is conceptual as well as case-based. The transformation of cities is better understood in historic and evolutionary perspectives. Empirically, we illustrate the transformation by analysing growth patterns at the city (municipal) level in Denmark. We look into specific urban experience-based strategies and projects in different countries. We pinpoint their global environmental implications and their dependency on expensive infrastructures. Finally, we discuss the entrepreneurial governance forms related to experience-based urban development.

In her article, Lorentzen (2009) discusses the spatial dimensions of the experience economy, with a particular view to the position of cities in spatial development. From an evolutionary perspective, she argues that the experience economy can be understood as a particular techno-economic paradigm with characteristics that are fundamentally different from those of the knowledge and the industrial economy. While the knowledge economy tends to concentrate activities and population in the metropolises because of the particular requirements for research and knowledge networks, the mechanisms of the experience economy is different. In place-bound experience products and services, leisure consumption is the driver, and urban amenities, cultural and sportive activities, rural and costal qualities, etc. therefore become essential. Big cities represent a varied and highly specialized offer of experience products and services, whereas more peripheral places may be attractive due to other characteristics: their authenticity, their natural environment or particular events or activities that are unique. In the development of an experience economy, cities need to provide low as well as highly skilled labour; they need to be accessible and visible—through branding or reputation—on the global market place. Mobility resources are crucial, both at individual and societal levels. New governance forms result from the need for new and flexible networks, including artists and fiery souls, big and small firms and public authorities. The question is whether the conditions related to experience-based development represent a window of opportunity for minor cities?

In spite of the fact that, lately, many cities and municipalities have embarked on experience-based strategies, no evidence has so far been provided that thoroughly assesses the potentials of the experience economy as an alternative basis of growth. Smidt-Jensen *et al.* (2009) set out to provide such evidence in their article. Using register-based employment data, the authors produce an image of the actual location dynamics of the experience economy in Denmark. In their analysis, the authors exclude the detachable and foot-loose sectors, which are claimed to agglomerate in big cities or clusters. Instead, focus is on the attendance-based experience industries that are bound to specific places. The notion of attendance-based experience industries is operationalized into the NACE categories of hotels and restaurants and entertainment, culture and sports. The authors find that, in general, the attendance-based experience industries in Denmark grew more than 30% from 1993 to 2006, covering 5.6% of employment that year. The growth in hotels and restaurants has concentrated in the big municipalities dominated by the big cities, whereas the growth in entertainment, culture and sport is more evenly distributed geographically. There is, however, a continued concentration of employment related to both hotels and restaurants and entertainment, culture and sport in two major cities as well as in the traditional tourist areas of Denmark. This does not mean that the growth of attendance-based experience industries in small and medium municipalities is negligible, but is not enough to bridge the gap. In terms of qualifications, the big cities take the lead, as their labour force in the experience sector is considerably highly educated. The evidence presented by Smidt-Jensen *et al.* (2009) implies that, in relation to urban growth policies, in the majority of the Danish municipalities, experience-based strategies can hardly stand-alone as an economic strategy.

The importance of experience-based projects and initiatives can be assessed along other lines than employment. They can also be discussed in terms of physical and spatial transformations in the city. Such transformations are partly due to the increased pressure on cities to brand themselves in order to improve their position in the global competition. In part, they can be seen as responses to internal challenges of the cities in relation to inclusion, leisure and learning. Marling *et al.* (2009) report from a project, in which they map out and interpret 15 Danish municipal cultural and learning projects. The projects include examples of buildings and urban architecture, performative urban spaces and urban events. The authors tentatively assess the projects with respect to their position in urban planning (e.g. lighthouse), their social characteristics and their function in the physical environment. The impression is that experience projects have become integrated in urban planning in Denmark, although such projects are quite heterogeneous. The Danish experience projects are either continuous (bazaars) or repeated (festivals). The projects may be strategic as part of overall urban planning or they may be the result of civic initiative, although not less influential than many municipal projects. Most of the projects studied are quite inclusive in relation to social groups, whom they offer entertainment as well as learning, the limit between which is, however, quite blurred. In sum, the projects can be seen as furthering a new urban political agenda and a new urban culture, in which the city invites to learning, experience and play in new transparent urban spaces and architectures.

Lassen *et al.* (2009) discuss what it takes to transform minor cities into spaces of experience. Transportation is shown to be a key ingredient in the construction of experience spaces. The production as well as the consumption of place-bound experiences is thus connected with high levels of mobility. The continued development of transport infrastructures and modes plays a significant role for people to be able to visit shopping centres, festivals and interesting places. At the local and regional levels, the use of cars plays a key role as means of transportation, whereas on the global scale, air transportation provides consumers with global mobility to visit places around the world as tourists. Based on the case studies from Sweden and Denmark, the authors identify new strategic alliances between local authorities in minor cities and global low-cost flight operators as key to the transformation of minor stagnating cities into vibrant experience spaces. Relative proximity to an airport represents a factor of location for young professionals wanting to combine urban careers and high mobility with attractive urban and natural environments for leisure. Proximity to an airport is also what enables experience businesses to achieve the critical mass of customers to attractions such as Legoland and related amenities. The environmental price of this strategy is, however, very high in terms of various kinds of pollution and green house gas emissions connected with air transportation. Until now, no means of (international, national or local) regulation or planning has yet been proved to be able to cope with environmental challenges arising from the air transportation that is deeply connected with the expanding global experience economy.

In his work on two German cities, Peter Allingham (2009) investigates the hypothesis that small or mid-sized cities in industrial and demographic decline may survive through strategies in which memorial qualities of the place is used to a considerable extent by planners. The means to achieve that have primarily been through storytelling and aesthetic appeal. Wolfsburg is a city with an industrial past in search of a new cultural profile, while Dresden is an old culture city in search of a more exciting profile. The two cities have invested in experiential strategies, in which the industrial past has been aestheticized, however, with communication and branding at the core of the strategies. This is not to be understood as merely a fanciful form of advertisement. The new branding strategies are more subtle in involving the audience in interactive rooms and frames within which the individual can unfold his/her project. The strategies are expected to turn expectations of potential investors in a more positive direction. Finally, in evaluating the branding methods applied in the two cities, the author refers to recent views on the question of representation and authenticity and the role of cultural heritage in experiential strategies.

The final article, by Therkildsen et al. (2009) discusses the relationship between experience-oriented development and urban governance and planning, based on a case study of the city of Frederikshavn (Denmark). The study shows how a varied range of experience-oriented projects have emerged and thrived in an entrepreneurial and more risk-taking new governance network that emerged in the aftermath of an extensive local economic crisis in 1999. Thereby, Frederikshavn seems to have managed to initiate a transformation process away from a declining mono-industrial city towards an economically more diversified city with a renewed identity. In investigating what influenced those changes, it is shown that municipal investments and internal reorganization and public–private cooperation played a significant role. Changes in networks, procedures and personnel were accompanied by the abolishment of traditional spatial (land use) planning. Instead, transformative urban growth strategies and more experimental and action-oriented approaches emerged. However, and finally, the authors also conclude that recent political tensions between growth and welfare agendas indicate that the case of Frederikshavn thereby exemplifies a test to the reaches or limits to government-supported neoliberal approaches in urban development and governance—and thereby also to the role of the local state.

References

Allingham, P. (2009) Experiential strategies for the survival of small cities in Europe, *European Planning Studies*, 17(6), pp. 905–923.
Andersson, Å. E. & Andersson, D. E. (2006) *The Economics of Experiences, the Arts and Entertainment* (Cheltenham: Edward Elgar).
Baltic Sea Solutions (2006) *Growth and Innovation Beyond Metropolises*. Holeby, Denmark; Palo Alto, USA.
Bell, D. & Jayne, M. (2006) Conceptualizing small cities, in: D. Bell & M. Jayne (Eds) *Small Cities: Urban Experience beyond the Metropolis*, pp. 1–18 (New York, NY: Routledge).

Bradley, A. & Hall, T. (2006) The festival phenomenon: Festivals, events and the promotion of small urban areas, in: D. Bell & M. Jayne (Eds) *Small Cities: Urban Experience beyond the Metropolis*, pp. 77–89 (New York, NY: Routledge).

Christgau, J. & Jacobsen, M. V. (2004) Byen i oplevelsessamfundet. Unpublished MA dissertation. Copenhagen, Copenhagen Business School.

Desrochers, P. (2001) Geographical proximity and the transmission of tacit knowledge, *The Review of Austrian Economics*, 14(1), pp. 25–46.

Florida, R. (2005) *Cities and the Creative Class* (London: Routledge).

Freestone, R. & Gibson, C. (2006) The cultural dimension of urban planning strategies: A historical perspective, in: J. Monclus & M. Guàrdia (Eds) *Culture, Urbanism and Planning*, pp. 21–41 (Aldershot: Ashgate).

Hajer, M. A. & Wagenaar, H. (Eds) (2003) *Deliberate Policy Analysis: Understanding Governance in the Network Society* (Cambridge: Cambridge University Press).

Healey, P. (2007) *Urban Complexity and Spatial Strategies Towards a Relational Planning for Our Times* (Oxon: Routledge).

Jacobs, J. (1985) *Cities and the Wealth of Nations. Principles of Economic Life* (New York: Vinatage Books).

Jantzen, C. (2007) Mellem nydelse og skuffelse. Et neurofysiologisk perspektiv på oplevelser, in: C. Jantzen & T. A. Rasmussen (Eds) *Oplevelsesøkonomi. Vinkler på forbrug*, pp. 135–163 (Aalborg: Aalborg Universitetsforlag).

Landry, C. (2006) *The Art of City Making* (London: Earthscan).

Lassen, C., Smink, C. & Smidt-Jensen, S. (2009) Experience spaces, (Aero)mobilities and environmental impacts, *European Planning Studies*, 17(6), pp. 887–903.

Løkke, J. (2006) Bybranding, fortælling og oplevelser. En undersøgelse af brandingens potentialer for den moderne provinsby eksemplificeret ved Horsens, Arhus Universitet, Århus.

Lorentzen, A. (2009) Cities in the experience economy, *European Planning Studies*, 17(6), pp. 829–845.

Lund, J. M., Nielsen, A. P., Goldschmidt, L. & Martinsen, T. (2005) *Følelsesfabrikken. Oplevelsesøkonomi på Dansk* (København: Børsens forlag).

Markusen, A. & Schrock, G. (2006) The distinctive city: Divergent patterns in growth, hierarchy and specialisation, *Urban Studies*, 43(8), pp. 1301–1323.

Marling, G., Jensen, O. B. & Kiib, H. (2009) The experience city: Planning of hybrid cultural projects, *European Planning Studies*, 17(6), pp. 863–885.

Mullins, P., Natalier, K., Smith, P. & Smeaton, B. (1999) Cities and consumption spaces, *Urban Affairs Review*, 35(1), pp. 44–71.

Nagy, E., Timar, J., Prömmel, J., Tille, D., Scheffler, N., Huttenloher, C. & Geser, G. (2007) *[Hist.Urban] Integrated Revitalization of Historical Towns to Promote a Polycentric and Sustainable Development*. Erkner (Germany): Institute for Regional Development and Structural Planning & Békéscaba (Hungary): Centre for Regional Studies, Hungarian Academy of Sciences.

Pine, J. B. II & Gilmore, H. H. (1998) Welcome to the experience economy, *Harvard Business Review*, July–August, pp. 97–103.

Pine, J. B. II & Gilmore, J. H (1999) *The Experience Economy* (Boston: Harvard Business School Press).

Porter, M. E. (1990) *The Competitive Advantage of Nations* (Hong Kong: The MacMillan Press Ltd).

Romein, A. (2005) The contribution of leisure and entertainment to the evolving polycentric urban network on regional scale: Towards a new research agenda, Paper for the 45th Congress of the European Regional Science Association, Free University of Amsterdam & European Regional Science Association, Amsterdam, August 23–27.

Schulze, G. (2005) *Die Erlebnisgesellschaft: Kultursoziologie der Gegenwart* (Frankfurt: Campus Verlag).

Smidt-Jensen, S., Skytt, C. B. & Winther, L. (2009) The geography of the experience economy in Denmark: Employment change and location dynamics in attendance-based experience industries, *European Planning Studies*, 17(6), pp. 847–862.

Therkildsen, H. P., Hansen, C. H. & Lorentzen, A. (2009) The experience economy and the transformation of urban governance and planning, *European Planning Studies*, 17(6), pp. 925–941.

Tofler, A. (1970) *Future Shock* (New York: Bentam Books).

Vetner, M. & Jantzen, C. (2007) Oplevelsen som identitetsmæssig konstituent. Oplevelsens socialpsykologiske struktur, in: C. Jantzen & T. A. Rasmussen (Eds) *Forbrugssituationer. Perspektiver på oplevelsesøkonomien*, pp. 27–55 (Aalborg: Aalborg Universitetsforlag).

Zukin, S. (1995) *The Cultures of Cities* (Cambridge, MA: Blackwell).

Cities in the Experience Economy

ANNE LORENTZEN

ABSTRACT *This article addresses the opportunities of cities, big and small, in the experience economy. It proposes an understanding of "experience economy", which encompasses not only entertainment and culture, but also services and places. To territorial development, the most interesting kind of experience consumption is the one co-located with its production. It is interesting because it invites people to stay and spend their money, either as residents or as tourists. Art and culture is known to cluster, and in big cities, the variety of the experience offer is an attraction in itself. Nevertheless, small cities embark on experience-based strategies, for example, related to events and branding. The article develops a theoretical framework that unfolds the territorial aspects of the experience economy. It does so in a comparative perspective, with a view to earlier (and coexisting paradigms), namely the industrial and the knowledge economy. Based on literature review it analyses the location patterns, the role of globalization, the changing governance forms and the mobility patterns, the latter being of utmost importance to the development of experience-based activities on the global market.*

Introduction

For decades, many cities have been suffering from job loss in the traditional industries, while they have had only little potential to attract knowledge-based activities, as these tend to concentrate in large metropolitan areas. A number of cities in Europe struggle against these trends by investing in tourism, attractions and activities and by branding themselves as hosts for global events. They do so with considerable success. What they do is to exploit the potentials of the experience economy. This contribution intends to lay the foundation for understanding the possibilities that the experience economy may offer in cities outside the centres of growth.

The article sets out to present an understanding of the experience economy that focuses on its place-bound dimensions. The article then develops an evolutionary approach to understand the particular spatial characteristics of the experience economy, and it suggests that the experience economy can be understood as a techno-economic paradigm with an economic geography which is different from foregoing paradigms. Hereafter, the article discusses what the experience economy may imply to the development of cities. It is argued that the experience economy represents a window of opportunity for cities that have had less favourable or even peripheral positions in other economic paradigms. Finally, the article discusses how this window of opportunity can be addressed by urban planning, and it is argued that there are important differences between what is known as culture planning and planning for the experience city.[1]

What Is Experience Economy?

Experience economy is a notion that intends to conceptualize a new trend in economic development, in which the driver is people's search for identity and involvement in an increasingly rich society. The notion is more encompassing than merely the market for entertainment and culture, or tourism, which are just aspects of the experience economy. The occupation with the cultural aspects of spatial or city development is a related new trend in urban geography, which tries to grasp the role of entertainment and culture in city development (Freestone & Gibson, 2006) and the impact of creative industries and culture economy (Scott, 2004, 2006). However, it is the business economists Pine and Gilmore (1998, 1999) that have become known for their notion of the "experience economy". The notion has gained a role not so much among private managers as it has among urban planners, among whom the notion has achieved the role of staging new discourses of urban development. In Denmark, several cities have embarked on experience projects, investing in festivals of the Vikings, the Middle Ages or Tordenskiold, or investing in experience-rich, harbour front environments, multi-purpose arenas or concert halls. The Danish government has even launched a programme with a focus on the experience economy (Regeringen, 2003).

At the outset, the notion of experience economy is related to a particular way to compete on the global market. To Pine and Gilmore (1998, 1999) an experience can be a competitive advantage of products. The experience economy is the latest stage of an evolution aimed at extracting as much value from the market as possible. From this perspective, the agrarian economy offered raw materials for the anonymous market, while the industrial economy offered manufactured standardized goods to the users. The service economy offers customized services to clients, while the experience economy offers personal experiences to the guests and customers (Pine & Gilmore, 1998, p. 98). Today, the success of a product, thus, depends on the experience that the product creates for the customer. Each stage represents a particular dynamic of value extraction, which is being outcompeted by the dynamic in the next stage.

It can be argued that entertainment businesses creating experiences is nothing new as we have known circuses and theatres for centuries and cinemas for decades. The new evolution is that today experiences become integrated in activities and products that were earlier trivial. Restaurants organize their services around particular themes (eatertainment). Shops and malls organize shows, events or expositions (shoppertainment). Mundane services like transportation are also being connected with experiences. Truly, the possibility of having individual experiences has always existed. However, with new technology the scope has considerably broadened. For example, interactivity can be built into many services and products, and the ability to create experiences for customers has become a value for the company. The experience content increases the value of services or goods on the market. In sum, the core of the experience economy is the still more generalized strategy of economic actors to "capitalize on experiences".

The Notion of Experience Products

According to Pine and Gilmore (1998, p. 98) "an experience occurs when a company intentionally uses services as the stage, and goods as props, to engage individual customers in a way that creates a memorable event". The experience derives from the interaction between the staged event and the individual's state of mind. Therefore, the experience is basically individual, although many individuals may have comparable experiences.

An experience can be considered a "product", since it must be produced or staged to be made available. Experiences are connected to the "consumption" of goods and services, by using them, by participating in activities and events or by visiting or living in the vicinity of places and attractions. Experience products are, thus, very varied and can be consumed in different ways. The common denominator of the experience products is the particular "relation" between the customer and the experience product (Pine & Gilmore, 1999). The customer may participate passively, as when enjoying a movie, or actively, as participant in activities such as hiking or attending a football match as supporter. Furthermore, the customer may simply absorb an experience being presented for him/her, or the customer may be part of the experience as participant. There are many combinations possible in the staging of experiences (Pine & Gilmore, 1998, p. 103). Experiences, thus, result from the design, development and introduction to the market. With reference to marketing research, it can be argued that in this process the quality of the involvement and even co-production of the consumer is important, as viewed by Bateson (2002) in relation to services. In this relationship, marketing the consumption of the service product is seen as a process rather than as the consumption of an outcome. Therefore, a capability of the experience producers is to be able to manage the prolonged relationships with the customers carefully (Grönroos, 2004). According to Grönroos (2004, p. 103), this means to "know the value system of the consumer that guides the need-fulfilling and value generating process" consumption.

The notion of "experience" can be qualified in different ways. For example, some of the experience products have the experience as its core (theatre). Other products have the experience as an add-on to known products (cell phones with game facilities). Some experience products have a high experience value, while others have a low such value. High or low value refers to criteria of novelty, repetition, unpredictability and personal involvement. This differentiation has been developed by Lund et al. (2005) and can be illustrated as an "experience compass" (Figure 1).

Experiences are enjoyable. This means that they create emotions in the person experiencing it. The emotional stimulation of experiences can be said to represent a stimulus for the individual to seek them (Jantzen, 2007; Vetner & Jantzen, 2007). This again may lead to identity formation at the individual as well as at the group level. This explains the drive towards experience consumption in today's "hedonistic" society (Schulze, 2005).

Pine and Gilmore suggest that the value of experiences is connected to the personal memory of experiences people have already had. This means that it is only an experience, when it is being told. The demand for and use of experiences is part of the creation of "identity", because the particular experiences used by someone are signals of the person someone wants to be (Lund et al., 2005). The value of the experience from this perspective is connected to the experience as "communication and narrative". This also means that the experience need not be unique. The visit to particular places is often repeated. Shopping as entertainment is neither unique nor very personal. It is connected to daily life experiences and becomes part of the identity of persons and groups of persons.

In Danish language there is a useful distinction between experience understood as a result of learning ("erfaring") and experience understood as a thrilling experience, as for example an adventure or an artistic experience ("oplevelse"). The latter is the focus of the experience economy, while the former is focused upon in the learning economy or in the knowledge economy (Lundvall, 1998).

From Experience Products to Experience Economy

It can be stated that not all experiences are marketed and that many experiences are free. Such experiences are not experience products and they are not part of the economy. Examples could be religious ceremonies, which are part of social institutions (churches) other than the market. Even marketed experiences are not new as a phenomenon. It is neither the existence of experiences in society as such, nor the production of experiences for the market, that is new. I suggest that the basic difference between then and now is the "structural context and role" of experiences in consumption, employment and production.

The profitable trade with experiences requires a mass market. A mass market is established through two types of economic development. One is the growth of income among large population groups, particularly on the northern part of the globe, which allows many people to spend money for luxury consumption. Today, large segments of people are wealthy enough to focus on consumption beyond the lowest levels of the Maslow pyramid of needs. They are willing to pay for the highest level of consumption, which is self-realization (Maslow, 1970). Self-realization is an activity that individuals undertake because they like it, not because they need to. Already in 1970, Alvin Tofler foresaw a situation in which the economy would be geared towards the provision of "psychic gratification" (Tofler, 1970). These activities are most often connected to leisure time. Andersson and Andersson (2006, p. 45) show how available leisure time has grown in the OECD countries. Rich people may be willing to pay higher prices for experience-rich products than they would have paid for trivial products. Disposable incomes have grown in the OECD countries, leading to an increasing share of the consumption dedicated for leisure activities, due to the high elasticity of luxury consumption (Andersson & Andersson, 2006, p. 42; see also Lorentzen & Hansen, 2009).

While material well-being is one of the pillars of the experience economy, the sociology of the modern society seems to be another. The emergence of the modern society means that people plan their lives individually and that there is an orientation towards the future. People orchestrate their future lives and, thus, their identities by planning to purchase particular experiences (Lund et al., 2005). Together, the two mega-trends of the economic and time surplus as well as the earlier-mentioned role of hedonism and identity creation represent the structural foundations of the experience economy.

In a more narrow sense, the reason why it is justified to talk about the emergence of an experience "economy" and not just about the growth of specific experience branches is, according to Pine and Gilmore (1999), that the experience strategy has relevance as a strategy of innovation and marketing for most branches. The reason for this is that all goods and services can be staged, and new experience stages can be designed. It is not primarily about what is sold (the content of the product or service that defines the experience-based strategy of businesses), but much more about the way it is sold through involving the customer. This view contrasts with the use of the word experience economy to cover the so-called creative industries, e.g. as in the policy of the Danish government (Regeringen, 2003). Similarly, the branches producing culture-related products and branches representing a high degree of creativity have been in focus in research (Scott, 2004, 2006). The creative branches include tourism, fashion, visual arts, radio/television, publishing firms, toys/entertainment, sports, architecture, design, film/video, advertising, edutainment, events, television, film, computer games and cultural institutions. However, the creativity, innovativeness and culture "content" of the products do not make them experience products. To be an experience product, there must be a certain "relationship" between the customer and the product. The drinking of milk from a friendly cow named Martha is an enjoyable and identity-creating experience just like attending an exciting theatre play is. Simply put, the focus of the creative economy discourse is on the innovativeness of the products. The focus of the experience economy discourse is on the customer, on his/her expectations and involvement with the product, and sometimes even as co-producer. In sum, the experience economy involves the growth of leisure consumption and the development of customer relationships.

Place Bound Versus Footloose Production

It can be argued that in the experience economy, "place" has a particular role to play. Places are, for example, often part of the identity connected with products and services. Examples abound in food, fashion and design. However, places are also being produced as a something in themselves, as when nice squares or parks are constructed and launched as places of sociability or recreation, shopping or entertainment. Places are, thus, being produced and marketed—they are being capitalized upon.

In comparison, culture industries and creative industries are footloose. Their products can be produced in any place where general conditions of production are fulfilled. They can also be sold to customers on the global market without any particular requirement to the location of consumption. In terms of location requirement, there is hardly any difference between the cultural and creative industries and any other knowledge-based industrial activity. They need a density of creative and innovative people.

The consumption of experience products is often place-bound, and so is much of its production. If we return to the experience compass (Figure 1), many place-bound experience products can be found in the upper left field (theme restaurants, regional gastronomy, theme parks and spectacular museums of art, performances and events). Such examples represent experience productions with high experience values and are, at the same time, pure experience products. New branches seem to struggle to get access to the upper left field of the experience compass. For example, in Denmark local banks have started to stage the visit of customers as a more holistic experience with drinks, "living room design" and sociability. In the opposite lower right field, we find products with a low experience value, which are products with the experience as an "add-on". Here we find more trivial products and services, where the experience is a bonus or an extra gift on top of the purchase. Many products have, for example, narratives printed on the packing. Likewise, mobile phones provide entertainment on top of their use as telephones.

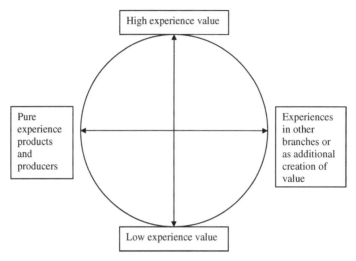

Figure 1. The experience compass (Lund *et al.*, 2005)

Such products and services are seldom dependent on the location of their production or consumption. They are "footloose".

Place-bound experience products can be divided into the following categories.

"Events" such as music festivals, historic festivals, sporting events, lectures by highly prestigious people are place-bound in a double sense. The final production of the event (organization, the playing of music, speaking) takes place in a particular place, an arena. A number of famous music festivals such as the Salzburg Festspiele (Austria) or the Roskilde Festival (Denmark) are certainly place-bound. In order to consume such events as a full experience, the customer has to be present in the same arena where the event takes place.

"Activities" such as shopping, hiking, participation in sport, handicraft production or artistic activities are on offer on particular locations. The location hosts such activities because of particular characteristics attached to the location, e.g. its history, local traditions, competences or the quality of the build or natural environment. In order to participate in the activities, the consumer has to be present at the location where the experience is being offered.

"Services" involving experiences include, for example, the serving of particular foods in theme restaurants, wellness services, exhibitions and art galleries, performances in theatres and cinemas. In Copenhagen, "Madelaine's food theatre" offers a complex combination of food and show. The food, design, light and performance of the staff produce experiences for all senses. The final production of the experience service takes place at particular locations and they have to be consumed at the same location in real time.

"Places" are more than just containers or "stages" for the production and consumption of experience services or goods. Places can also be seen as experience products in themselves, or they may constitute parts of experience products. Places are here understood as the built and the natural environment on different scales. The production of places involves the physical planning, construction and maintenance of buildings and natural resorts. It also involves the creation of attention, or "branding" of places. Communication is a very important part of the production of experience places (Frandsen *et al.*, 2005). It involves the production of the place in people's minds. Places like castles, squares, parks, woods, beaches, malls, museums, as well as parts of towns or whole cities need to be made positively known for their particular attractions. In order to make people relate to the place, planned communication processes are required to support the development and enhancement of the place brand, quite in parallel to relationship marketing of services (Grönroos, 2004, p. 102).

Tentatively, it is suggested here that "the role of place in the experience economy is to increase the experience value of the products on the market. The place constituent increases the experience value by means of identity creation and the involvement of the consumer". The relationship between place and experience value is illustrated in Figure 2.

The important role of place in the production of pure and high-value experience products does not mean that the production is always limited to one particular place. The production of an experience product may be planned in one place, developed in another and staged for consumption in a third (O'Dell, 2005). The production chain may, thus, be geographically dispersed, such as in relation to the development and planning of major events involving international artists, the preparation of food in a restaurant or the exhibition of art in an art gallery. It is the final production and the consumption of the experience product that needs to be co-located. In this way, place-bound experience products have similarities with services that are being produced in the same time as they are consumed. The customer, therefore, in many cases needs to be present at the place of the service production in order to consume it, even if some services can be consumed at a distance by means of communication technologies. In the marketing of experience products, the role of place can be seen when providers try to create a wish of the consumer to "be there". In sum, experience products are often place-bound, and the role of place is to increase the value of the product. In the following, some examples of pure high-value and place-bound experience productions will be considered.

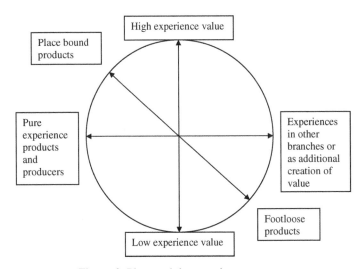

Figure 2. Place and the experience compass

In discussing the role of places in the experience economy, O'Dell (2005) suggests the notion of "experiencescape", understood as places in which experiences are being staged and consumed. The notion is somewhat similar to Pine and Gilmore's conceptualization of "stage". In addition to an understanding of the place as stage, it is suggested here that "place as such is even being consumed" by tourists (Urry, 1995) and citizens (Glaeser, 2001). Places are not only spaces of entertainment and pleasure, and meeting grounds of diverse groups with overlapping or conflicting interests, as O'Dell (2005) suggests. Places are also consumed by people enjoying the atmosphere, the sociability and even the identity that can be created by being present. Places are strategically planned, laid out and designed by urban planners to meet such a demand for atmosphere and sociability. A subgroup of experiencescapes is "nostalgiascapes" (Gyimóthy, 2005), which are places representing historic identity and lifestyles. The power of experiencescapes in producing feelings of identity is sometimes very strong. Special buildings connected to experiences may even help redefine the whole identity of cities. Well-known examples are the Guggenheim museum in Bilbao, which changed people's perception of the city, which again influenced economic growth in the region. This is now known as the "Guggenheim effect". Minor examples can be found in, for example, Århus, Denmark, where the House of Music had a comparable impact.

The close connection between place and experience production and consumption will be developed in more detail below, where the role of experience economy for local development will be discussed.

An Evolutionary Approach

The significance of the experience economy for local development and planning is best analysed in comparison with other types of economies. Pine and Gilmore's historic approach is somewhat sketchy and related only to the strategy of businesses. From a developmental perspective, it seems useful to embark on a more comprehensive approach as an outset for discussion. For this purpose, the notion of techno-economic paradigms introduced by Perez (1985) is fruitful. According to Perez, drawing upon Schumpeter (1939) and Kondratiev (1935) among many others, the economy develops in long waves, each characterized by particular key factors, such as coal, steam power, oil or microelectronics. It is characterized by specific cost structures, investment patterns, location geographies and inter-branch relationships. Each wave is also characterized by particular socio-institutional frameworks, involving relative proportions and character of public and private responsibility, the provision of education and training, the distribution of income, the organization of workers and major interest groups, among other things. The paradigms and institutions evolve historically, but coexist for long periods of time. Each paradigm has a life cycle, beginning with the generation of innovations, which are diffused in the economy. This is the point of departure for prolonged periods of growth, based on high profits and increasing productivity. Limits to growth are found when applications are fully exploited and incremental innovations have been developed to the frontier. The next wave will take its point of departure in a new innovation (Perez, 1985, p. 443). Today, this general approach is recognized by many and called evolutionary and institutional economy (Perez, 2004). In the context of this article, it is particularly interesting how Perez suggests that each techno-economic paradigm has its own "specific geography". Places may, therefore, lose or win with the change of techno-economic paradigms.

When developing the geographic characteristics of the experience economy and in particular its implications for the development of cities, the notion of techno-economic paradigms represents a promising approach, which can be used to develop an understanding of
the logic and dynamics of the experience economy, and in particular, its implications for
local development. This choice of approach does not imply, however, that the experience
economy is actually suggested as a well-defined techno-economic paradigm. Analysed as
a techno-economic paradigm, the experience economy is connected with the production
and consumption of experiences, as defined above. Experiences are becoming increasingly
constitutive in the economy, as people use still more money for experience-based consumption. In Denmark, expenses related to leisure, communication and transport, thus,
constituted 22% of the consumption in 2003 (Danmarks Statistik, 2007). Investments in
experience-based production and landscapes are equally growing. New locations
emerge as economically dynamic places of experiences, while old (industrial) locations
offer new attractions. The change in geography is supported by the still lower costs for
transport and communication, which enables high mobility, and high mobility can be considered a key ingredient of the experience economy. Information and communication technologies (ICT) is a key technology in the experience economy, but rather than defining
technology-related innovations as the starting point for the techno-economic paradigm
of the experience economy, a bundle of socio-economic trends serve as basis for it, of
which increasing affluence, increasing mobility and the globalization of competition of
products (Dicken, 2003) and places (Brenner, 2004) are most important.

At the institutional level, the experience economy also offers new developments. New
actors, networks and interdependencies emerge, resulting in new strategic alliances promoting the production and consumption of experiences. New areas in education emerge
focusing on e.g. tourism, leisure and culture management (Kultur og Kommunikation,
2007): e.g. today, most Danish universities offer experience related Master's programmes.

In terms of geography, the experience economy rather than the knowledge economy
seems to be less amenable to concentrating in large metropolitan areas. Part of the experience production and consumption is place-bound, as we have seen. Places representing
and hosting experiences are, at a first glance, quite diverse. Experience products, activities
and places are, thus, connected to places like villages, beaches, mountains and cities of
different sizes and histories.

Urban Development and Techno-economic Paradigms

Urban development results, among other factors, from the location of economic activities.
It is, therefore, relevant to consider the role and function of cities in different techno-economic paradigms. From the perspective of location and urban development, it seems justified to consider three principally distinctive paradigms of development: the industrial, the
knowledge and the experience-based paradigm. In the following, these paradigms will be
discussed in terms of space.

In the industrial paradigm, firms have clustered in cities in order to reduce costs by being
proximate to markets, supplies and labour (Hayter, 1998). In terms of consumption, people
located in cities because they wanted to socialize and benefit from different facilities and
services (Glaeser, 2001). In the knowledge economy, firms tend to locate in cities with
access to specialists, research and decision-making centres, preferably in very large metropolises (Simmie, 2003). The specialists and the creative classes have tended to locate in

large cities with a varied offer of culture, and interesting jobs (Florida, 2002, 2005; Scott, 2006). The metropolises have grown to the detriment of smaller cities, reinforced by the intensified competition among cities (Brenner, 2004; Simmie, 2003; van den Berg *et al.*, 2004, 2005). In the experience economy, on the contrary, small cities and peripheral places have begun to produce events, places, activities related to culture, heritage and authenticity (Bell & Jayne, 2006; Bradley & Hall, 2006; Meethan, 1996; Wilks-Heeg & North, 2004). In Denmark, there are numerous examples of small cities gaining new roles based on events (e.g. annual Rock Festivals in Skanderborg, Nibe, Ringe) and innovative branding (e.g. Horsens and Frederikshavn) (Frandsen *et al.*, 2005; Løkke, 2006). The implication is that in the experience economy, the location of people and economic activity may potentially de-concentrate as the centres of gravity become redefined. The forces of globalization, which have reinforced the concentration of people and economic activities during the industrial- and knowledge-based paradigms, seem to support de-concentration in the experience economy when peripheral places become integrated in the global flow of people, money and information. From a perspective of place, quality smallness may even represent an attraction (Bell & Jayne, 2006). Table 1 tentatively contrasts the economic geographies and dynamics of the three economic paradigms.

Table 1. Industrial location and techno-economic paradigms

Dimension	Techno-economic paradigm		
	Industrial economy	Knowledge economy	Experience economy
Production/ location	Concentration in advanced regions	Concentration in metropolises of the advanced regions	Many locations in central and peripheral countries and regions
Consumption/ location	Concentration in advanced regions	Concentrated in metropolitan areas	Attractive places (big and small) in developed and less developed places
Globalization	Separation and dispersion of production	Flow of knowledge, goods, people and capital between the metropolises of the advanced regions	Integration of different experience places in the global flow of information, people and money
	International trade Direct investment		
Role of the centre	Advanced industrial production and services	Knowledge production	Magnet of inhabitants and visitors
	Research and development	Research and development	Big and specialized offer of experiences based on variety and history
	Decision-making	Decision-making	
Role of the periphery	Raw materials Simple industries	Simple industries Global services	Tourism-based growth Experiences based on authenticity and natural environment
	Low-cost labour		Activities, events

Table 1 illustrates how economic geography development in the experience economy may be characterized by patterns and dynamics that are different from those of earlier techno-economic paradigms. This is because in the experience economy it is possible to capitalize on a greater variety of places and the resources attached to them. Experience consumers are mainly people from highly developed places. Their experience consumption is not geographically restricted. Both central and more peripheral places are, thus, places of consumption. Globalization integrates all sorts of locations in the global flow of information, people and money in the experience economy. Central places are magnets of inhabitants and tourists because they offer a wide array of experience products. The variety of experiences is attractive to customers, but the specialization in them may also attract many customers. Peripheral places cannot provide a wide diversity of experiences, but may offer instead particular experiences rooted in authenticity and natural environments. Experience-based growth in the periphery is likely to be based on tourist flows or on the inflow of additional population, e.g. retired people such as in southern Spain and in small coastal towns in Denmark.

Planning for the Experience City

The city is increasingly seen as a social space in which cultural activities and events can develop. The city is also seen as a place, a built and a natural environment, which can be made attractive to citizens, visitors and firms. In an interesting and attractive city, traditional economic activities are also likely to develop, and to nourish from the creative atmosphere (Lund *et al.*, 2005).

A functional approach is helpful to understand the increased focus on culture and leisure in urban planning. It can be shown how city functions change along with the change in techno-economic paradigms. A distinction between different economic paradigms in relation to urban development has been suggested by Hall (2001), who sees a development from an "industrial economy" towards an "informational economy" and today a "cultural economy". The three paradigms may correspond grossly with the three paradigms illustrated in Table 1. There are resemblances between the "knowledge economy" and Hall's "informational economy", but not between the "experience economy" and Hall's "culture economy". According to Hall, in the cultural economy, the role of the city is to provide advances in transport and communication, quality in residential and environmental terms and high levels of cultural and educational offering. Particularly important in urban interventions is to make "strategic use of cultural resources". There are many examples of cultural planning aiming at local economic development. It has critically been noted that there need not be any link between the strategic use of culture assets on the one hand, and culture policy on the other (Vaz & Jaques, 2006, p. 243). Culture and capital have been linked strategically, leading to what Harvey (2000) calls a commodification of the city. In comparison, in earlier planning paradigms, culture has been in focus in its own right, not as a means of economic development. Cities have been seen as works of art, or culture flagships have been developed as part of "welfare policies" (Freestone & Gibson, 2006). Cultural planning, in the strategic sense, has been recorded since the mid-1970s in big cities of Europe and North America. Its rationale has been to create attractive locations for individual and collective consumption. This again is seen as a precondition for the attraction of wealthy people to visit towns and reside in them, and also for the attraction of investment and production (Evans, 2001).

It can be argued that "cultural planning" for local development is connected, not to any specific "cultural economy", but to the techno-economic paradigms of late industrialism, and particularly to that of the knowledge economy. The connection between the knowledge economy and culture has been described by Florida. Florida (2002, 2005) provides documentation supporting the close relationship between the location of high-skilled labour, technological innovation in industry and growth on the one hand, and a varied supply of culture on the other (Florida, 2002, 2005). Following the Florida hypothesis, Clark (2004) discusses the increasing focus on urban cultural amenities as a lever for urban growth.

Evans (2001, p. 141) seems to approach a more detailed understanding of the city in the experience economy by making a distinction between different approaches and functions to leisure and work in cities. Evans suggests five different approaches. The "culture city" focuses on museums, galleries, theatres and concert halls. The "historic city" offers museums and monuments, including the offers of the culture city. The "night life city" offers theatres, concert halls, night clubs, red-light districts, cafes and restaurants. The "shopping city" offers cafes, restaurants, shops and offices, while the "tourist city" has it all. Those five city types may inspire the search for "the experience city". Each of the "cities" represents aspects of the place-bound consumption of pure experiences (Figure 2). The cities invite citizens and visitors to enjoy plays, to identify with a historical past, to eat typical food and to shop for typical products. However, it must be noted that Evans' "model cities" need to be relatively big, diverse and centrally located to fulfil these functions. The role of city size is supported by Andersson and Andersson (2006), who point at the agglomeration economies in cultural production and the role of clustering in artistic production. The cities in focus are "culture capitals", world cities and metropolitan areas (Scott, 2004, 2006). The big city bias seems to be due to the application of a conventional concept of culture. Against the big city bias speaks the fact that culture consumption and production constitutes only part of experience consumption and production. Experience cities need not be based on traditional culture provision. They may have many other attractions to offer.

This means that cities need not be big to be attractive for leisure and tourism. Centralization is not a precondition for the development of all sorts of experience production and consumption. It means that both small and big cities can be part of the experience economy. An "experience city" need not be fully equipped with amenities. The important characteristic of an experience city is its attractive atmosphere, which comes from place-bound activities, events and services, attractive places and diverse social spaces, which make visitors and residents feel inspired, involved and connected to the place. Such characteristics imply that the city need not be big. Experience city atmosphere can be developed in smaller cities as well (Bell & Jayne, 2006).

Experience-Based City Growth

The geographic characteristics of the experience economy make it attractive for urban developers to consider in connection with strategies for urban growth and development in the periphery. They may consider experience growth a "window of opportunity" for small cities and cities in industrial decline. But what are the preconditions of success for strategies of experience-based urban growth? Is it possible to develop a more structured approach to the question, taking a point of departure in the paradigmatic perspective on economic development?

To urban planners it would be relevant to identify the potential of the city to attract and develop experience-based consumption and production. Here, the three techno-economic paradigms will be applied. Each of the paradigms can be briefly described in relation to the factors of location (requirements) of firms and people and their possible correspondence with the actual conditions of location (resources, amenities) (Hayter, 1998). In this article, the location considered is the city. The requirements that cities fulfil in the three paradigms can be characterized in functional and economic terms. It is, however, also relevant to consider the particular mobilities and the different forms of governance that characterize the economic dynamic of the three paradigms.

The role of the city for production is different in the three paradigms. The labour pool, transport node and provision of physical facilities characterize cities in the industrial economy, while the city must be able to provide highly skilled labour, information and communication networks and proximity to knowledge providers in the knowledge economy. The amenities and resources needed in the city in the experience economy can be deduced from the preceding sections. Tentatively it can be suggested that in the experience economy the city must provide a wide array of labour skills to be able to produce experiences. The city must be accessible for outsiders and exposed on the market by branding and information, so that the attractions become known. The resources needed for the production of experiences are related to the history, culture and nature and inherited competences of the city.

Knowledge is also a key resource in the experience economy. For instance, in order to produce a rock concert, it involves high skills concerning logistics, acoustics and lighting as well as artistic skills of the object of the concert. Information and advertising activities require knowledge of communication and media. The whole infrastructure of the event needs to be produced. Here, practical skills are needed related to construction, maintenance, organization, parking services, the serving of customers and cleaning, just to mention some. Relational skills and artistic creativity are also required.

There is a major difference between the roles of the city in the knowledge economy compared with the experience economy. While few cities provide resources sufficient for knowledge-based growth, many cities provide the basis for the production of marketable experiences. Small cities may have few special cultural and natural resources on the basis of which they can produce attractive experience products, while big cities may develop a bigger variety as well as a more specialized offer. The labour demand of the experience economy fits with the labour supply of small cities, where low-skilled labour often abounds. Thus, there seems to be potentials for experience-based production in both small and big cities.

"Consumption" in the industrial economy is based on large scales and low prices. The industrial city provides supermarkets, cheap housing and welfare services and infrastructure. In the knowledge economy, consumption is more individualized, and the requirements for urban quality and entertainment increase due to the demands of the creative class, which large cities compete to attract. In the experience economy, consumption becomes a driver of growth, and urban quality development becomes the means to attract consumers. People consume sociability, partake in activities and develop identities and individualities based on their urban living. Both visitors and residents are part of the experience space of the city. They make the place interesting and sociable. Different cities are attractive to consumers, not because of size but because of their individual qualities. Both small and big cities are potentially able to attract citizens and tourists.

The three paradigms are different in terms of "connectivity" and economic and social "mobility". The industrial economy is characterized by localized, material-intensive and labour-intensive activities in relatively fixed places. In the knowledge economy, knowledge is produced and shared in fluid spaces among experts and scientists working globally, while the location of activities often depend more on aesthetic than on functional qualities connected to the city. The experience economy is connected with high individual mobility. People choose their residence due to individual quality criteria and commute to their job if necessary. People are also willing to travel far to consume experience products such as theme parks or festivals in weekends and during holidays. Places are constructed through branding, placing them on the global market. Branding concerns the way the city competes with other localities to attract consumers by creating and communicating narratives about its attractions. This phenomenon has been labelled "the catwalk economy" (Löfgren, 2003). It has been studied how this task of communication has been carried out in the Danish city of Horsens (Frandsen *et al.*, 2005). Horsens used to be known for its state prison. Today, it is known for its concerts with international stars. The change of the city narrative of the city of Frederikshavn has been studied by Lorentzen (2007). This city changed its reputation as a violent and sad city in industrial decline with high levels of unemployment into a creative provincial environment, and even an experience city. In this way, cities embark on strategies of "glocalization" (Brenner, 2004).

"Mobility" can, thus, be seen as an important basis for the development of experience-based development in cities. Infrastructure and strategic communication (in terms of "branding") create the experience city. This implies that remote and little-known places may have a disadvantage in the experience economy.

In addition, "actors and governance" structures differ between the three paradigms. While the industrial economy is characterized by stable, hierarchic structures in firms and planning, clear-cut authorities, and the state as provider of welfare, the knowledge economy is connected with the dissolution of such structures. Network structures among firms and authorities develop, resulting in complex interdependencies in which the state is facilitating private initiatives more flexibly (Healey, 2007). In the experience economy, many actors emerge in the development and production of experience products. They include small service producing firms, networks of firms, multinational corporations, urban planning authorities, civic organizations and fiery souls (Hjorth & Kostera, 2007; Lorentzen, 2008). The variety of experience products means that they are developed and produced in different kinds of networks. For example, unique events like a rock show require temporary and often extra-local networks, while the production of attractive tourist accommodation may be integrated in stable global production chains. In any case, networks are important in experience production because of the need to source different types of knowledge to produce and market innovative and unique experiences. While big cities have the advantage of being well networked globally, small cities may have the advantage of small distance and many bonds between the local actors (Lorentzen, 2008). Small cities, thus, have a particular ease of networking which is beneficial for the creation of experiences.

The ideas presented in this section are summarized in Table 2. The table organizes the key requirements to cities in each of the three techno-economic paradigms, and in doing so it distinguishes between the requirements related to production, to consumption, to mobility and to governance.

Table 2. The role of cities in techno-economic paradigms

Dimension	Techno-economic paradigm		
	Industrial economy	Knowledge economy	Experience economy
Production	Labour pool	Pool of skilled labour	Supply of low- and high-skilled labour
	Transport node	Information and communication network	Accessibility
	Physical facilities	Proximity to universities Research institutions	Exposure, branding History, authenticity
Consumption	Mass consumption	Individual consumption	Sociability, activity, identity, individuality
	Stores and supermarkets	Attractive malls	Experience products, services, places
	Transport of persons	Attractive housing	
	Cheap housing	Leisure space	
	Welfare services	Culture supply	
Mobility	Place (fixed)	Space (fluid)	Construction ("branding") of places
	Functional qualities	Functional and aesthetic qualities	Accessibility
			High individual mobility Consumption-based identities
Governance	Hierarchy	Network	New entrepreneurial forms and networks
	Authority	Polycentric	Temporary networks
	Mono-centric	Tailored to context	Changing stakeholders
	Welfare	Facilitation	

Conclusion

The experience economy is a notion that originates in business economics and seems useful to denote the new trend in economic development in which the driver is leisure consumption. Today, places compete on the "global catwalk" to attract citizens, tourists and firms and they do so by developing their qualities. Place-bound experience production and consumption seems to be of particular interest to local development because of the implications for employment and for the quality of place. Based on an evolutionary approach, the article develops the geographic perspectives of the experience economy and compares the location requirements of this economy with those of earlier techno-economic paradigms. It is argued that in the experience economy, the location requirements are such that central as well as peripheral places, small as well as big cities, may embark on experience economic development. A high degree of concentration of consumption in big cities provides the critical mass for the development of a large variety of specialized experience products related to culture, shopping and dining, while small cities and peripheral places build their experience production on the local culture and nature resources. A common requirement for all places embarking on experience economic growth and development is that they are accessible and known, or in other words that they are integrated in the global flow of people and information. New actors are involved in the development of experiences where civic initiative may produce experience events. The field of actors in the experience economy is dynamic, and the networks often temporary. The governance is often quite blurred, as voluntary, public and commercial experience projects evolve and merge, making the place interesting and attractive.

Because of those characteristics it could be argued that the experience economy may represent a "window of opportunity" for small cities that tend to be marginalized in the knowledge economy. Small cities have an opportunity because they hold resources such as natural beauty, historic urban cores, traditional production, tranquillity, intimacy, local networks and civic initiative, and because of the enabling of global transportation and communication possibilities. This does not mean, however, that the magnetism of big cities no longer represents a threat to small city development. It only means that small cities may find niches in the global demand for experiences.

Note

1. The article is based on a paper presented on the Regional Studies conference "Regions in Focus" 2–5 April 2007 in Lisbon, and is a contribution to the research programme on the City in the Experience Economy at the Department of Development and Planning, Aalborg University, Denmark.

References

Andersson, Å. E. & Andersson, D. E. (2006) *The Economics of Experiences, the Arts and Entertainment* (Cheltenham: Edward Elgar).

Bateson, J. (2002) Consumer performance and quality in services, *Managing Service Quality*, 12(4), pp. 206–209.

Bell, D. & Jayne, M. (2006) Conceptualizing small cities, in: D. Bell & M. Jayne (Eds) *Small Cities: Urban Experience Beyond the Metropolis*, pp. 1–18 (New York, NY: Routledge).

van den Berg, L., Pol, P. M. J. & van Winden, W. (2004) *Cities in the Knowledge Economy* (Rotterdam: The European Institute for Comparative Urban Research (EURICUR)).

van den Berg, L., Pol, P. M. J., van Winden, W. & Woets, P. (2005) *European Cities in the Knowledge Economy*, 1st edn (Aldershot: Ashgate).

Bradley, A. & Hall, T. (2006) The festival phenomenon: Festivals, events and the promotion of small urban areas, in: D. Bell & M. Jayne (Eds) *Small Cities: Urban Experience Beyond the Metropolis*, pp. 77–89 (New York, NY: Routledge).

Brenner, N. (2004) Urban governance and the production of new state spaces in Western Europe 1960–2000, *Review of International Political Economy*, 11(3), pp. 447–488.

Clark, T. N. (2004) *The City as an Entertainment Machine*, 1st edn (Oxford: Elsevier).

Danmarks Statistik (2007) *Statistisk årbog 2006* (København: Danmarks Statistik).

Dicken, P. (2003) *Global Shift*, 4th edn (London: Sage).

Evans, G. (2001) *Cultural Planning: An Urban Renaissance?* 1st edn (London: Routledge).

Florida, R. (2002) *The Creative Class* (New York, NY: Basic Books).

Florida, R. (2005) *Cities and the Creative Class* (London: Routledge).

Frandsen, F., Olsen, L. B., Amstrup, J. O. & Sørensen, C. (2005) *Den kommunikerende kommune*, 1st edn (København: Børsens Forlag).

Freestone, R. & Gibson, C. (2006) The cultural dimension of urban planning strategies: A historical perspective, in: J. Monclus & M. Guàrdia (Eds) *Culture, Urbanism and Planning*, pp. 21–41 (Aldershot: Ashgate).

Glaeser, E. K. J. S. A. (2001) Consumer city, *Journal of Economic Geography*, 2001(1), pp. 27–50.

Grönroos, C. (2004) The relationship marketing process: Communication, interaction, dialogue, value, *Journal of Business and Industrial Marketing*, 19(2), pp. 99–113.

Gyimóthy, S. (2005) *Nostalgiascapes* (Copenhagen: Copenhagen Business School Press).

Hall, P. (2001) *Cities in Civilisation* (New York, NY: Fromm International).

Harvey, D. (2000) *Spaces of Hope* (Edinburg: Edinburg University Press).

Hayter, R. (1998) *The Dynamics of Industrial Location: The Factory, the Firm and the Production System* (West Sussex: John Wiley).

Healey, P. (2007) *Urban Complexity and Spatial Strategies* (Oxon: Routledge).

Hjorth, D. & Kostera, M. (2007) *Entrepreneurship and the Experience Economy*, 1st edn (Copenhagen: Copenhagen Business School Press).

Jantzen, C. (2007) Mellem nydelse og skuffelse. Et neurofysiologisk perspektiv på oplevelser, in: C. Jantzen & T. A. Rasmussen (Eds) *Oplevelsesøkonomi. Vinkler på forbrug*, pp. 135–163 (Aalborg: Aalborg Universitetsforlag).

Kondratiev, N. D. (1935) The long waves of economic life, *Review of Economic Statistics*, 17(6), pp. 105–115.

Kultur Og Kommunikation (2007) *Creative Industries Education in the Nordic Countries* (Oslo, Norway: Nordic Innovation Centre).

Löfgren, O. (2003) The new economy: A cultural history, *Global Networks*, 3(3), pp. 239–253.

Løkke, J. (2006) *Bybranding, fortælling og oplevelser. En undersøgelse af brandingens potentialer for den moderne provinsby eksemplificeret ved Horsens* (Århus: Århus Universitet).

Lorentzen, A. (2007) *Frederikshavn indtager "the global catwalk"* (Aalborg: Institut for Samfundsudvikling og Planlægning).

Lorentzen, A. (2008) *Knowledge Networks in the Experience Economy: An Analysis of Four Flagship Projects in Frederikshavn*, 1st edn (Aalborg: Department of Development and Planning).

Lorentzen, A. & Hansen, C. J. (2009) The role and transformation of the city in the experience economy: Identifying and exploring research challenges, *European Planning Studies*, 17(6), pp. 817–827.

Lund, J. M., Nielsen, A. P., Goldsmith, L. & Martinsen, T. (2005) *Følelsesfabrikken. Oplevelsesøkonomi på dansk*, 1st edn (København: Børsens forlag).

Lundvall, B.-Å. (1998) The learning economy: Challenges to economic theory and policy, in: K. Nielsen & B. Johnson (Eds) *Institutions and Economic Change*, (1st edn), pp. 33–54 (Cheltenham: Edward Elgar).

Maslow, A. H. (1970) *Motivation and Personality*, 2nd edn (New York, NY: Harper and Row).

Meethan, K. (1996) Consuming (in) the civilized city, *Annals of Tourism Research*, 23(2), pp. 322–340.

O'Dell, T. (2005) Experiencescapes: Blurring borders and testing connections, in: T. O'Dell & P. Billing (Eds) *Experiencescapes. Tourism, Culture, and Economy*, pp. 11–33 (Køge: Copenhagen Business School Press).

Perez, C. (1985) Microelectronics, long waves and world structural change: New perspectives for developing countries, *World Development*, 13(3), pp. 441–463.

Perez, C. (2004) Technological revolutions, paradigm shifts and socio-institutional change, in: E. Reinert (Ed.) *Globalization, Economic Development and Inequality: An Alternative Perspective. New Horizons in Institutional and Evolutionary Economics*, pp. 217–242 (Cheltenham: Edward Elgar).

Pine, J. B., II & Gilmore, J. H. (1998) Welcome to the experience economy, *Harvard Business Review*, 76(4), pp. 97–103.

Pine, J. B., II & Gilmore, J. H. (1999) *The Experience Economy* (Boston, MA: Harvard Business School Press).

Regeringen (2003) *Danmark i kultur og oplevelsesøkonomien—5 nye skridt på vejen. Vækst med vilje* (København: Regeringen).

Schulze, G. (2005) *Die Erlebnisgesellschaf*, pp. 1–589 (Frankfurt/New York: Campus Verlag).

Schumpeter, J. A. (1939) *A Theoretical, Historical and Statistical Analysis of the Capitalist Process* (New York, NY: McGraw-Hill).

Scott, A. J. (2004) Cultural-products industries and urban economic development: Prospects for growth and market contestation in global context, *Urban Affairs Review*, 39(4), pp. 461–490.

Scott, A. J. (2006) Creative cities: Conceptual issues and policy questions, *Journal of Urban Affairs*, 28(1), pp. 1–17.

Simmie, J. (2003) Innovation and urban regions as national and international nodes for the transfer and sharing of knowledge, *Regional Studies*, 37(6&7), pp. 607–620.

Tofler, A. (1970) *Future Shock*, 1st edn (New York, NY: Bentam Boook).

Urry, J. (1995) *Consuming Places*, 1st edn (London: Routledge).

Vaz, L. F. & Jaques, P. B. (2006) Contemporary urban spectacularisation, in: J. Monclus & M. Guàrdia (Eds) *Culture, Urbanism and Planning*, pp. 241–253 (Aldershot: Ashgate).

Vetner, M. & Jantzen, C. (2007) Oplevelsen som identitetsmæssig konstituent. Oplevelsens socialpsykologiske struktur, in: C. Jantzen & T. A. Rasmussen (Eds) *Forbrugssituationer. Perspektiver på oplevelsesøkonomien*, pp. 27–55 (Aalborg: Aalborg Universitetsforlag).

Wilks-Heeg, S. & North, P. (2004) Cultural policy and urban regeneration: A special edition of local economy, *Local Economy*, 19(4), pp. 305–311.

The Geography of the Experience Economy in Denmark: Employment Change and Location Dynamics in Attendance-based Experience Industries

SØREN SMIDT-JENSEN, CHRISTINE BENNA SKYTT &
LARS WINTHER

ABSTRACT *Recently the "experience economy" has been promoted as a vehicle for urban and regional growth, also in peripheral cities and regions. Little evidence is, however, provided to sustain this claim. To inform the discussion of the experience economy as a potential for urban and regional growth, the article provides an analysis of location dynamics and employment growth of a specific segment of the experience economy, the attendance-based experience industries, in Danish municipalities from 1993 to 2006. Based on the analysis, it is concluded that the emerging experience economy in the Danish context produces new forms of uneven geographies: first, employment growth is significantly higher in large cities compared with that in small- and medium-sized cities, and second, the level of education for persons employed in the experience economy is higher in the largest cities compared with that in small- and medium-sized cities. Hence, the potential of the experience economy as a vehicle for growth even in peripheral cities and regions has in many cases (not yet) been fulfilled. Thus, using the experience economy as a lever to obtain future prosperity may be a very fragile strategy for the majority of cities and municipalities outside the main growth centres and classic tourist destinations.*

1. Introduction

According to Amin and Thrift (2002, p. 70) and Pine and Gilmore (1999, p. 2), increasing competition in the market means that "goods and services are no longer enough" and that producers must differentiate their products by transforming them into "experiences" which engage the consumer. The same process is arguably affecting cities and regions worldwide

31

as they brand themselves into experiences for residents and visitors alike (Richards, 2001). Much of the "experience creation" that is presently taking place is driven by an aspiration of public authorities to develop the productive resources of their regions or cities, as traditional sources of income and job creation decline. In recent years, policy-makers and planners throughout Scandinavia have embraced the rise of "the experience economy" as they believe it to bring new perspectives for urban and regional development, also in peripheral cities and regions (Sørensen *et al.*, 2007; Sundbo & Bærenholdt, 2007; Vaekstfonden, 2007; Regeringen, 2003; Manniche & Jensen, 2006; Haraldsen *et al.*, 2004; Mossberg, 2003; KK Stiftelsen, 2003; Lassen *et al.*, 2009). It seems that the idea of the experience economy has gained a stronghold in Scandinavia and, in particular, in the Danish debate on urban and regional development—although elsewhere in Europe and North America there is plenty of evidence of similar policies which seek to promote industries and growth-based experiences under various terms and conceptualizations (Bell & Jayne, 2006; Clark *et al.*, 2004; Fiore *et al.*, 2007; Dammers & Keiner, 2006; Roberts & Hall, 2004). The basis for such strategies is, however, often rather weak and undocumented. The core belief in many strategies is that experiences can become integrated in all economic activities, also the ones that were earlier seen as trivial. The experience factor gives economic actors an advantage when their products (including places) are brought on to the market. Recently, some support for the idea that the experience economy brings new perspectives for local development and growth can be found in recent research papers, although often based on a set of case studies (Fuglsang *et al.*, 2008; Sørensen *et al.*, 2007; Manniche & Jensen, 2006; Bærenholdt & Haldrup, 2006, Therkildsen *et al.*, 2009).

In reality, the economic geography of the experience economy and its potential for urban and regional economic growth are somewhat understudied. Most of the academic literature evolves around conceptualizations of the term and there is little empirical evidence of its economic importance, spatial growth patterns, job creation and quality, its demand for human capital, its processes of innovation and knowledge production and the impact on urban and regional growth outside the main growth corridors. Furthermore, current research on the experience economy has so far been less preoccupied with making the concept operational for empirical analysis.

The first purpose of this study is to make an operational definition of the experience economy to provide empirical evidence of its local and regional impact in terms of jobs. The second purpose of this study is to contribute to the discussion of the experience economy as a vehicle for urban and regional growth and provide empirical evidence of the spatial dynamics of the experience economy in Denmark. Using register-based employment data, this study provides an analysis of the location dynamics of a specific segment of the experience economy, the attendance-based experience industries, in Danish municipalities from 1993 to 2006. Based on the analysis, we explore the geography that is being produced by the recent growth of the experience economy. This is done in order to provide a basic understanding of the spatial dynamics of the experience economy and to inform a discussion of the claimed potential that it can be a driver of urban and regional growth, also outside the main city regions. We add another dimension to get one step deeper into the economic geography of the experience economy: we analyse the labour qualifications in terms of formal education to examine the kind of jobs that have been created. Is it predominately "McJobs" signifying low skilled, low wages and little job training or "iMacJobs" that require considerable training and involve high skills, wages and autonomy (Allan *et al.*,

2006; Lindsay & McQuaid, 2004)? The result provides an indication of what kind of growth the experience economy provides.

To do so, the article is organized as follows. The next section moves beyond conceptualization and makes an operational definition of the experience economy. A distinction is made between detachable experience products and services (DEPS) and attendance-based experience products and services (AEPS). The section concludes with a description of the data used to analyse the economic geographies of the experience economy in Denmark. Before the specific analysis of the industries, the third section of the article sketches out the dynamics of the economic geography of Denmark. The section further analyses the geography of AEPS in Denmark, its location patterns, employment growth and human capital (using educational data), and discusses the differences between various sub-sectors of AEPS. The last section discusses the results of the analysis in relation to the claims embedded in academic and political arguments concerning the potential of the experience economy as an engine of economic growth in cities and regions outside major city regions. The analysis will provide a new basis for this discussion and inform the debate on the experience economy as a strategy and vehicle for local and regional growth.

2. Defining the Experience Economy

Pine and Gilmore (1999) take "the experience" beyond the provision of goods and services to the recognition of experience as a distinct economic offering. Their basic premise is that experiences represent an existing but previously unarticulated genre of economic output that have the potential to distinguish business offerings. As an economic offering, experiences can add value to a business's goods and services and are distinct from both. The customer who buys a service buys a set of intangible activities carried out on his/her behalf. The purchase of an experience, on the other hand, buys time enjoying a series of memorable events that engage the consumer in a personal way (Pine & Gilmore, 1999). Accordingly, to Pine and Gilmore (1999, p. 4), "the experience economy" first and foremost refers to a new evolution in the way that products are sold. They argue that economic actors gain an advantage in the market by staging and selling memorable experiences that are enjoyable and personally engaging for the customer. This is claimed to be a vital part of the future of capitalist economies (Thrift, 2005, p. 7). In this context, the experience economy does not refer to a particular industry or a specific segment of the economy. It is understood as a general and qualitatively new characteristic of advanced capitalism, and it would make no sense to estimate the experience economy's magnitude or significance in terms of generated revenues or employment. It is a way for firms to increase their profits and reduce the uncertainty in the market. The definition proposed by Pine and Gilmore (1999, p. 6) sees the experience economy as a general process in the economy and accordingly experiences can become integrated in all economic activities, also the ones that were earlier seen as trivial. Since the experience component of a product or service is increasingly becoming the basis for profit, and because an experience component in theory can be added to all products and services (Sundbo & Bærenholdt, 2007), the recipe to follow for the economic actors, cities and regions is, in a nutshell, "just add experience". This new economic evolution is argued to be based on the fact that customers are generally becoming more open to experiences and more willing to take greater risks and spend larger sums of money to experience something new, because basic goods and services no longer satisfy customers sufficiently (O'Dell & Billing, 2005; Mossberg, 2003; Amin & Thrift, 2002).

Another way of defining the experience economy, however, refers to specific industries where the experience component is particularly strong or where the potential for developing the experience component is evident (Mommaas, 2004). By using such definitions, the experience economy easily comes to embrace cultural industries (Beck, 2003; Hesmondhalgh, 2002; Pratt, 1997), creative industries (Caves, 2000; Fleming, 1999), leisure industries (Roberts, 2004), entertainment industries (King & Sayre, 2003; Hannigan, 1998) and tourism (Weiermair & Mathies, 2004; Ioannides & Debbage, 1998). The way the term is used in the current public debate seems to blend the two definitions and thus leave the term very diverse and blurred: it seems to include everything from (street) musicians to sleeping bag manufacturers and art museums—and many claim a membership to the experience economy. Thus, the experience economy is a characteristic of advanced capitalism, but it is more evident in certain specific sectors of the economy. What sectors and industries in the economy can be characterized as experience economy?

2.1. *The Primary Sector of the Experience Economy*

Sundbo & Bærenholdt (2007, p. 11) suggest to make a divide between two sectors of the experience economy: a primary experience sector that is composed of companies and institutions that have the production of experiences as the primary objective and a secondary experience sector where experiences are add-ons to artefacts or services, e.g. games in mobile phones, coffee sold with a certain brand and story attached to it (see also Lund, 2005). Using this division of the experience economy, we suggest a further division of the primary experience sector into two sub-categories: producers of DEPS and producers of AEPS.

Characteristically, DEPS can be transmitted to consumers all over the world either as artefacts or as electronic/digital impulses (e.g. music CDs, MP3 files, broadcasting, books, etc.). These products can be sent off to receivers everywhere and utilized whenever a consumer wants to. Usually, products within this category are easily reproducible. Thus, the original site of creation is perhaps in a music studio in London with creative musicians and technicians but the fabrication of the product takes place elsewhere as a part of the global division of labour. This implies that the consumption, and partly the production of DEPS, has a rather "footloose" nature at least in terms of relations to a specific place. It is evident from the studies of many of the creative and cultural industries that these footloose sectors tend to agglomerate in the big cities or clusters (Scott & Power, 2004; Pratt, 2005).

The second sub-category, AEPS, which is the sector of the experience economy that this article examines more in depth, covers economic activities where the majority of the production of a certain experience product or service is bound to a specific physical place, and at the same time, the consumption of the experience requires *in situ* attendance in a physical place. Usually, this will require that the consumer is moving outside his/her home to enjoy the product (e.g. to a restaurant, cinema, sports arena, etc.). In some instances, the co-presence of the consumer and the producer/performer/creator in the same particular place will be a precondition for the realization of the experience (e.g. the guest and the chef in a restaurant, the fan and the rock band at a concert, etc.), while the creator in other instances is absent (e.g. a guest in an art museum watching a painting by Van Gogh—although a curator could be characterized as at least the co-creator of such an experience). Some AEPS are even detachable to some extent and can be set up in other places, either as permanent constructions (e.g. the Guggenheim franchises in Bilbao, Abu Dhabi and Guadalajara) or as shows that are place-bound, and are open for attendance only temporarily (e.g. a circus on tour). Importantly, however, an attachment to a particular physical place is a precondition for the consumption of AEPS. Such physical places where AEPS can be consumed have been termed "experiencescapes"—for an elaboration (O'Dell & Billing, 2005, p. 15).

2.2. *Data*

In the next section of the article, we analyse the location and geography of AEPS in Denmark using register-based employment data at municipality level.[1] We will provide an analysis of the location of two broad NACE (Nomenclature generale des Activités economiques dans les Communautes Européennes) categories that best cover AEPS: "NACE 55 Hotels and Restaurants"[2] and "NACE 92 Entertainment, Culture and Sport".[3] We focus on those two categories because the majority of the products and services in both groups of industries require that the consumer is present to consume the experience and together they provide a strong overview over the industries and hence the firms of AEPS. We are well aware that with modern technology, entertainment such as concerts and football games can be experienced in other places than where they actually occur, and they can be displaced in time. However, our focus is on employment and regional growth which constitute the supply side of the experience economy and its potential for urban and regional growth, and not the direct consumption. The years of 1993, 2002 and 2006 have been selected. In 1993, the latest period of growth of the Danish economy took off, while 2002 signifies a high employment growth, and 2006 is the latest year available. The data from 1993 to 2006 have been established according to the former administrative structure in Denmark; in 2006, the number of Danish municipalities was cut down from 275 to 98. Employing those distinctions makes it possible not only to compare the data over time but also to get a more nuanced and detailed picture of the Danish geography.

A broad definition of experience products and services is employed; it includes entertainment, amusement parks, hotels, restaurants, etc. These industries have been through a change from traditional services to providers of services as experiences. The choice is made for two reasons. The first is pragmatic: to make a comprehensive analysis on the basis of the data that are available to us, it has been necessary to use the relatively high level of industry aggregation to make an analysis on the level of municipalities. Second, it seems that many urban and regional policies include it all. The economic geographies of these industries are seldom formerly described in the literature on urban and regional growth. Before we turn to the specific analysis of these industries, we will briefly outline the dynamics and main characteristics of the economic geography of Denmark in the last decades.

3. The Economic Geography of the Experience Economy in Denmark

There has been an economic resurgence of the large cities and their surrounding urban landscapes in Europe and also in Denmark (Winther & Hansen, 2006; Hansen & Winther, 2007; Hanell & Neubauer, 2005; Hall & Pain, 2006). A long-term period of growth of the Danish economy in terms of production and jobs started in the early

1990s, and although the millennium crisis of the new economy reduced employment and slowed down the growth, the economy has regained most of the losses in recent years even though more gloomy economic times seem to be approaching as we write. Concurrently, with the economic growth, a new economic geography has emerged. First, the new growth pattern reveals a marked concentration of job growth in the large cities of Denmark. Especially, the capital Copenhagen and the second largest city Aarhus have revitalized. Secondly, the growth of the large cities is not just a resurgence of the central city but also a growth of the surrounding suburbs and even the outer city, a transformation zone between the city region and countryside (Winther & Hansen, 2006; Hansen & Winther, 2007). Accordingly, new urban landscapes have emerged, and in the case of Copenhagen, it covers most of Zealand (Andersen *et al.*, 2006). Thirdly, outside the main urban landscapes of Denmark, cities and municipalities are struggling with recession, stagnation, job loss, economic and industrial restructuring, deindustrialization, declining population, brain drain and unemployment (Miljøministeriet, 2006; Engelstoft *et al.*, 2006). Even the success cities of 1970s and 1980s are struggling, and many seek to maintain the previous success by turning to new activities in tourism and to the experience economy (Smidt-Jensen, 2008). This is not only a crisis in remote islands or small fishing communities on the west coast of Jutland, but a challenge to even medium-sized cities located just outside the two main urban growth regions. Thus, small- and medium-sized cities face different challenges, depending on their position in the Danish space economy. Small- and medium-sized cities inside the city regions have a different potential for economic success than cities outside the main city regions.

Figure 1 shows one central take on the Danish space economy in 2006 in terms of employment. It indicates the main four cities in Denmark: Copenhagen, Århus, Aalborg and Odense. It is obvious that there is a marked concentration of employment in and around the four main cities and on the east coast of Jutland, together with a relatively small number of medium-sized cities/municipalities outside the dominant city regions. This strong, city-oriented economic geography has been reinforced since the 1990s. Table 1 demonstrates that from 1993 to 2006 the general employment growth was strongest in the largest dominant Danish municipalities (8.4%) in the central urban landscapes. This growth is well above the national average (5.7%). The lowest growth was found in small municipalities (2.4%). The recent growth patterns are different, however. After the millennium crisis, the small and largest municipalities lost employment while the medium-sized municipalities had zero growth. The next section analyses the Danish geography of AEPS and frames it within the general economic landscape of Denmark.

3.1. *The Geography of AEPS in Denmark*

In 2006, AEPS employed 142,722 people in Denmark, accounting for 5.2% of the total Danish employment. The growth of AEPS has been strong since the 1990s, indicating the increasing importance of the experience economy in the Danish economy. The employment in the sector grew with more than 30% from 1993 to 2006. Employment growth was strong in both main NACE categories: hotels and restaurants and entertainment, culture and sports. Table 2 shows the employment growth from 1993 to 2006, and both hotels and restaurants and entertainment, culture and sports had impressive growth rates well above the national average. Hotels and restaurants grew with 25.4% and entertainment, culture and sports with 37.3%. Contrary to the Danish economy as a whole, the growth in AEPS took off

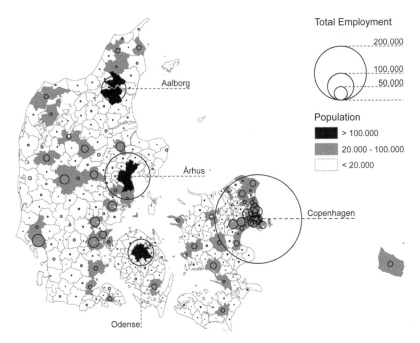

Figure 1. Employment in Denmark (2006)

with significant growth rates in the period 2002–2006—especially the four largest municipalities have dominated this development. AEPS grew with 44.3% in the large municipalities
which is well above national growth, while the growth was more modest but still above the
general employment growth in the small municipalities (17.9%). There are, however, significant differences between the two NACE categories. The growth of employment in hotels and
restaurants has been very strong in the large municipalities from 1993 to 2006, with a growth
of more than 52% compared with a mere 9.1% growth in small municipalities. The medium-
sized municipalities have a growth just below the average national growth. The employment
growth from 1993 to 2006 in entertainment, culture and sports was more evenly distributed
between the three categories of municipalities analysed. The growth was even higher and
above the national average in the small municipalities, while growth in medium-sized

Table 1. Employment growth in Denmark (1993–2006)

	Growth	
Employment, all branches	1993–2006	2002–2006
Small municipalities, <20,000 inhabitants	2.5	−1.6
Medium-sized municipalities, 20,000–100,000 inhabitants	6.6	−0.1
Large municipalities, <100,000 inhabitants	8.4	−1.2
Denmark	5.7	−0.9

Source: Statistic Denmark and own calculations.
Note: Small municipalities: 209 municipalities in 2006; medium-sized municipalities: 62 municipalities in 2006;
large municipalities: 4 municipalities in 2006.

Table 2. Employment growth in entertainment, culture and sports and hotels and restaurants in Denmark (1993–2006)

	Growth	
	1993–2006	2002–2006
AEPS (NACE 55+NACE 92)		
Small municipalities, <20,000 inhabitants	17.9	11.8
Medium-sized municipalities, 20,000–100,000 inhabitants	27.1	16.8
Large municipalities, <100,000 inhabitants	44.3	36.0
Denmark	30.2	21.8
Hotels and restaurants (NACE 55)		
Small municipalities, <20,000 inhabitants	9.1	4.9
Medium-sized municipalities, 20,000–100,000 inhabitants	20.7	10.8
Large municipalities, <100,000 inhabitants	52.8	40.6
Denmark	25.8	17.4
Entertainment, culture and sports (NACE 92)		
Small municipalities, <20,000 inhabitants	42.4	29.9
Medium-sized municipalities, 20,000–100,000 inhabitants	37.3	26.4
Large municipalities, <100,000 inhabitants	34.9	30.6
Denmark	37.3	28.9

Source: Statistics Denmark and own calculations.

municipalities was around national average. It was just below the national average in the large municipalities. The share of employed persons of the total number of employed persons in entertainment, culture and sports in small municipalities was, however, very modest. In 2006, 1.4% of employed persons worked within entertainment, culture and sports, which was below the national average (2.0%) and much below the average for large municipalities (3.3%; see also Appendix 2). The employment growth in entertainment, culture and sports between 2002 and 2006 was almost identical between the categories of municipalities. It should be noted that entertainment, culture and sports also contains employment in film, video, radio, television and press agencies where the level of education tends to be relatively high. These industries are usually located in the largest cities.

Figure 2 shows the location and geography of hotels and restaurants and entertainment, culture and sports and reveals several characteristics of AEPS in Denmark. Looking closer at entertainment, culture and sports (map A), the map shows that the relative importance of this industry in terms of the share of the local working population employed in entertainment, culture and sports is most important in the capital of Copenhagen and its surrounding municipalities, in the city of Odense (third largest city), and generally in traditional tourist destinations of which many are on the Danish coast. One of the "in-land" municipalities outside the Copenhagen region, where the share is high, is the municipality of Billund in South Jutland which hosts the amusement park Legoland; the most important tourist attraction outside Copenhagen. In absolute terms, Copenhagen has the largest entertainment, culture and sports sector followed by Aarhus, Odense and Aalborg (indicated on the map by the size of the circles). Taken together, this underlines that the high share of entertainment, culture and sports in the largest municipalities is mainly due to the high share in the municipality of Copenhagen. The map also reveals that there are large parts of Denmark where the entertainment, culture and sports industry is

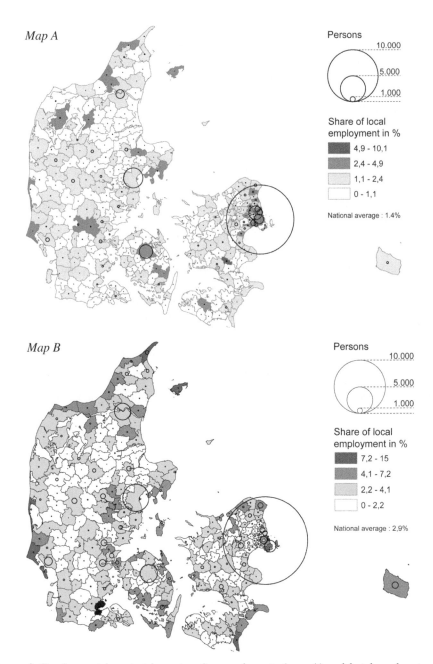

Figure 2. Employment in entertainment, culture and sports (map A) and hotels and restaurants (map B) in Denmark (2006)

relatively unimportant. This is the case in central parts of Jutland, central parts of south Jutland, central parts of Funen (with the exception of Odense) and most parts of Zealand outside the Copenhagen region. The municipalities in the latter areas are mostly small or medium-sized and in many cases located in-land.

A somewhat similar pattern can be detected on map B, which shows employment in hotels and restaurants. Again, the traditional tourist resorts turn out to have the highest relative share of employment in hotels and restaurants (e.g. the municipalities on the west coast of South and North Jutland). In all the four large cities, the relative importance is on the national level, with the municipality of Kastrup just outside central Copenhagen having a share that is above the national average (due to the Copenhagen Airport in Kastrup which has a high number of employed persons in restaurants, catering and other supply firms located just around the airport). As was the case for entertainment, culture and sports, there are parts of Denmark where the importance of employment in hotels and restaurants is very modest. This is the case for large parts of central Jutland, central Funen, western Zealand and south Zealand. The patterns have some correspondence with employment in the two AEPS industries: where employment in entertainment, culture and sports is relatively high and employment in hotels and restaurants tends to be relatively high as well, and vice versa, although there are exceptions like the municipality of Skagen (the tip of Jutland) where the share of people employed in entertainment, culture and sports is lower than the national average, while the share in hotels and restaurants is among the highest in the country.

So far, we have concerned ourselves mainly with the spatial distribution of the experience economy, but what kinds of jobs are being produced? The level of education in AEPS provides an indication of this.

3.2. AEPS Employment and Labour Qualifications

In this section, we add an extra dimension to the economic geography of AEPS. We analyse the labour qualifications in terms of formal education, in order to examine what kind of jobs that have been created. The result provides an indication of the kind of growth the experience economy provides and what kind of jobs are created and where.

Table 3 shows the labour qualifications in terms of the formal education involved in the two AEPS industries in relation to all branches. The numbers indicate differences in education level between the two AEPS categories and also differences in the level of education in the three groups of municipalities. In hotels and restaurants, the level of education compared with that in all other branches is very low. A very large share of the employed in hotels and restaurants only has primary school as their formal educational background, while only about 3% has a university degree. In comparison, employees in the entertainment, culture and sports have a higher educational background, also compared with the industry average in Denmark. Even though the share of employees in entertainment, culture and sports with primary school background is close to the national average, the share of employed with a university degree in entertainment, culture and sports is much higher than the national industry average (12.6% compared with 8.6%). Considering the numbers for the different groups of municipalities, it shows that the employees in the two AEPS industries (as in all industries) are more qualified in the largest municipalities compared with those in medium and small municipalities. With respect to employment in hotels and restaurants in the small municipalities, almost every second employee has primary school as the highest formal educational background, while in the largest cities it is only about one-third. In line with this, the share of employees in hotels and restaurants with a university degree is higher in the largest municipalities and lower in small- and medium-sized municipalities. In entertainment, culture and sports, it is remarkable that

Table 3. Level of education of employed persons in hotels and restaurants and entertainment, culture and sports in Denmark (2006)

	Group 1	Group 2	Group 3	Not specified	Total
Hotels and restaurants					
Small municipalities	48.1	43.9	2.9	5.0	100.0
Medium-sized municipalities	46.9	46.2	1.8	5.1	100.0
Large municipalities	33.5	54.4	4.9	7.2	100.0
Denmark	42.9	48.2	3.2	5.8	100.0
Entertainment, culture and sports					
Small municipalities	30.7	56.1	9.9	3.3	100.0
Medium-sized municipalities	28.4	59.4	9.5	2.6	100.0
Large municipalities	19.4	60.4	17.3	2.9	100.0
Denmark	25.5	59.0	12.6	2.9	100.0
All branches					
Small municipalities	30.8	62.9	3.8	2.5	100.0
Medium-sized municipalities	24.4	65.7	7.8	2.1	100.0
Large municipalities	18.7	63.3	15.5	2.5	100.0
Denmark	24.9	64.1	8.6	2.3	100.0

Source: Statistics Denmark and own calculations.

Note: Group 1: primary school; Group 2: high school and college degree; Group 3: university degree.

the level of education is much higher in the largest cities (17.3%) compared with that in medium-sized and small cities (9.5% and 9.9%).

This pattern indicates that in the part of AEPS industries that consist of hotels and restaurants, it is not highly educated employees that occupy the jobs, especially not outside the main growth areas of Denmark. In entertainment, culture and sports, on the contrary, the AEPS workforce with a university degree tends to be a significant share of the total, especially in the largest municipalities where more than 17% of employees are holding a university degree, but also in small- and medium-sized municipalities where almost 10% has a university degree. The numbers also underline that the two AEPS industries, which we have concentrated upon, are very different in terms of labour qualifications.

4. Discussion and Concluding Remarks: The Experience Economy as Potential for Urban and Regional Development?

In current debates concerning the experience economy in relation to urban and regional studies, much emphasis has been placed on conceptualization of the term and how it may contribute to urban and regional development. It has been argued from planners, politicians and even researchers that the experience economy provides new growth opportunities for small- and medium-sized cities, even in peripheral regions and cities. One of the main problems of using the experience economy as a strategy to promote urban and regional growth is that it is not yet fully understood how small- and medium-sized cities develop, including how the experience economy contributes to current urban growth. It has been unclear what the fortunes and fates might be for those places under such changing conditions and in particular where it leaves small- and medium-sized cities outside the large city regions (Bell & Jayne, 2006). One reason for this is that the economic geography of the experience economy has been understudied in terms of spatial growth patterns, job creation and quality, its demands of human capital and its impact on urban and regional growth.

Furthermore, the phrase "the experience economy" has been used in so many ways and contexts that it has become blurred and imprecise. To overcome those shortages and to inform the discussions of the experience economy and urban and regional development, we have presented an operational definition of the experience economy. Here, we suggested to divide the primary experience sector, e.g. the sector that is composed of firms and institutions that have the production of experiences as their primary objective, into two sub-categories: (1) AEPS and (2) DEPS. The focus of this article is on the economic geography of AEPS, specifically on employment growth, location patterns and the use of human capital.

The empirical evidence revealed that employment growth of AEPS has been very strong in the Danish economy. The analysis showed that the employment in the AEPS grew more than 30% from 1993 to 2006. In particular, the growth took off between 2002 and 2006. This emphasizes the importance of the experience economy in terms of jobs and job creation in the Danish economy, especially when comparing with the general growth rates. There are, however, significant differences between the two AEPS sub-sectors. The employment growth in hotels and restaurants is very unevenly distributed with very high growth rates in the largest municipalities and a more moderate growth in the medium- and small-sized municipalities, whereas the employment growth in entertainment, culture and sports is more evenly distributed between the Danish municipalities. It is important to notice that the distribution of the employment growth is different from hotels and restaurants—the employment growth in entertainment, culture and sports was highest in the small municipalities and lowest in the large municipalities.

The analysis of human capital (level of formal education) revealed that the two AEPS industries have different characteristics. Hotels and restaurants is certainly not a category dominated by knowledge workers, especially outside the main growth areas of Denmark, the share of employees with a university degree is remarkably low. On the other hand, the entertainment, culture and sports industries generally employ a relatively large share of highly educated persons. This is most significant in the largest municipalities.

The analysis also provides evidence of the local impact of hotels and restaurants and entertainment, culture and sports. The evidence supports that outside the main urban landscapes the experience economy possesses the potential to create growth in terms of jobs even in peripheral regions. The jobs created may not be for high-end knowledge workers, but they probably have a better match with local labour markets. However, the potential of the experience economy seems to be mostly evident in the traditional tourist destination in Denmark: for instance, in the municipalities along the coastal areas, and, in particular, municipalities that are favoured by a rich history, have a long-term tradition in entertainment or are evolved around a singular entrepreneurial spirit and project (e.g. the Legoland park in Billund or Bon Bon Land in Fladså). The remaining majority of municipalities have not been notably affected by the rise of the experience economy in the past decades.

Thus, the idea that the experience economy can be a window of opportunities for small- and medium-sized municipalities outside the growth centres is partly valid when it comes to the traditional tourist places with natural, cultural or entertainment amenities. However, for a majority of cities and municipalities, AEPS as a way to future prosperity may turn out to be a very fragile growth strategy, at least in terms of jobs and job creation. The experience economy is not the new normative panacea for regional or local economic ills that planners and planning can use if things are going bad, but it can, in some cases, be a fruitful path for prosperity if the necessary conditions are present.

Analytically, however, the experience economy provides a theoretical perspective that can contribute to understand the uneven economic geographies that are currently being produced, including answers as to why some cities and regions even outside the main growth regions have seen growth while others stagnated or even declined.

Furthermore, the analysis presented in this article has been limited to considering employment in terms of growth and human capital. This has proven to be of relevance in order to be able to question the experience economy as a policy perspective for urban and regional development. However, it is also evident that more knowledge and research is needed to be able to understand how and where the experience economy contributes to local and regional growth, and how it contributes to the production of uneven economic geographies. First, more empirical evidence and quantitative analysis are needed. Here, studies of economic growth in terms of turnover or GDP, productivity and local multiplier effects will prove useful. Secondly, a better understanding of the relational networks of the experience economy is needed, including those of ownership and control to evaluate the potentials of urban and regional growth outside the main growth regions, but also innovative and entrepreneurial networks of the experience economy. This will contribute to answer basic questions such as who will benefit from the experience economy in the future? It will also make it possible to examine whether there is a match between local policies and growth strategies and the economic reality of the experience economy.

Notes

1. The data are workplace based, meaning that the persons are registered according to the municipality in which they work.
2. NACE (Nomenclature generale des Activités economiques dans les Communautes Européennes) is the common nomenclature of economic sectors in the European Union. NACE 55 consists of hotel, conference centres, camp sites, holiday centres, restaurants, discotheques, cafés, canteens and catering.
3. NACE 92: Entertainment, film and video production, cinemas, television and radio broadcasting, performing arts (theatre, concerts), artists, management of cultural institutions, amusement parks, press agencies, libraries and archives (public and research), museums, botanical gardens, zoos, sports (sport clubs, sport facilities, marinas, lottery and gambling). Although NACE 92 comprises industries that cannot be categorized as AEPS (film and video production, television and radio broadcasting entertainment, film and video production, cinemas, television and radio broadcasting and press agencies), they are included because the majority of NACE 92 can be attributed to AEPS (approximately 80–85% of which 70–75% is located in the municipalities of the Greater Copenhagen Region).

References

Allan, C., Bamber, G. & Timo, N. (2006) Fast-food work: Are McJobs satisfying? *Employee Relations*, 28(5), pp. 402–420.

Amin, A. & Thrift, N. (2002) *Cities: Reimagining the Urban* (Cambridge: The Polity Press).

Andersen, H. T., Engelstoft, S., Møller-Jensen, L. & Winther, L. (2006) *Det danske bylandskab, Geografi*, vol. 2006, pp. 55–57, Annual Report, Royal Geographical Society.

Beck, A. (2003) *Understanding the Cultural Industries* (London: Routledge).

Bell, D. & Jayne, M. (Eds) (2006) *Small Cities: Urban Experience Beyond the Metropolis* (London: Routledge).

Bærenholdt, J. O. & Haldrup, M. (2006) Mobile networks and place making in cultural tourism, *European Urban and Regional Studies*, 9(3), pp. 209–224.

Caves, R. E. (2000) *Creative Industries: Contracts between Art and Commerce* (Cambridge, MA: Harvard University Press).

Clark, T. N., Lloyd, R., Wong, K. K. & Jain, P. (2004) Amenities drive urban growth: A new paradigm and policy linkages, in: T. N. Clark (Ed.) *Research in Urban Policy: Vol. 9. The City as an Entertainment Machine*, pp. 291–321 (Oxford: Elsevier).

Dammers, E. & Keiner, M. (2006) Rural development in Europe, *DISP*, 42(3), pp. 5–15.

Engelstoft, S., Jensen-Butler, C., Smith, I. & Winther, L. (2006) Industrial clusters in Denmark: Theory and empirical evidence, *Papers in Regional Science*, 85(1), pp. 73–97.

Fiore, A. M., Niehm, L., Oh, H., Jeong, M. & Hausafus, C. (2007) Experience economy strategies: Adding value to small rural businesses, *Journal of Extension*, 45(2), pp. 1–13, Article No. 2IAW4.

Fleming, T. (1999) *The Role of Creative Industries in Local and Regional Development* (Manchester: Government Office for Yorkshire and the Forum on Creative Industries).

Fuglsang, L., Højland, J., Sundbo, J. & Sørenseb, F. (2008) *Innovation i oplevelsesvirksomheder—Resultater fra en survey* (Roskilde: CEUS Centre for Leisure Management Research).

Hall, P. & Pain, K. (Eds) (2006) *The Polycentric Metropolis: Learning from Mega-City Regions in Europe* (London: Earthscan).

Hanell, T. & Neubauer, J. (2005) *Cities of the Baltic Sea Region—Development Trends at the Turn of the Millennium*, Nordregio Report 2005:1, Stockholm: Nordregio.

Hannigan, J. (1998) *Fantasy City: Pleasure and Profit in the Postmodern Metropolis* (London: Routledge).

Hansen, H. K. & Winther, L. (2007) The spaces of urban economic geographies: Industrial transformation in the outer city of Copenhagen, *Danish Journal of Geography*, 107(2), pp. 45–58.

Haraldsen, T., Flygind, S. K, Overåg, K. & Power, D. (2004) *Kartlegging av kulturnæringene i Norge—økonomisk betydning, vekst- og utviklingspotensial* ØF-apport nr. 10: 2004, Lillehammer: Østlandsforskning.

Hesmondhalgh, D. (2002) *The Cultural Industries* (London: Sage).

Ioannides, D. & Debbage, K. (1998) The economic geography and tourism nexus, in: D. Ioannides & K. Debbage (Eds) *The Economic Geography of the Tourist Industry: A Supply-Side Analysis* (London: Routledge).

King, C. & Sayre, S. (2003) *Entertainment and Society: Audiences, Trends and Impacts* (London: Sage).

Stiftelsen, K. K. (2003) *Upplevelsesindustrin* (Stockholm: KK Stiftelsen).

Lassen, C., Smink, C. & Smidt-Jensen, S. (2009) Experience spaces, (aero)mobilities and environmental impacts, *European Planning Studies*, 17(6), pp. 887–903.

Lindsay, C. & McQuaid, R. W. (2004) Avoiding the "McJobs". Unemployed Job Seekers and Attitudes to Service Work, *Work, Employment and Society*, 18(2), pp. 297–319.

Lund, J. (Ed.) (2005) *Følelsesfabrikken: oplevelsesøkonomi på dansk* (København: Børsens Forlag).

Manniche, J. & Jensen, A. B. (2006) *Kultur- og oplevelsesøkonomi i nordiske landdistrikter—Indhold og perspektiver* (København: Nordisk Ministerråd).

Miljøministeriet/Danish Ministry of the Environment (2006) *Landsplanredegørelse 2006. Det nye Danmarkskort—planlægning under nye vilkår* (København: Miljøministeriet/Ministry of the Environment).

Mommaas, H. (2004) Cultural clusters and the post-industial city: Towards the remapping of urban cultural policy, *Urban Studies*, 41(3), pp. 507–532.

Mossberg, L. (2003) *Att skapa upplevelser—från OK till WOW* (Lund: Studentlitteratur).

O'Dell, T. & Billing, P. (Eds) (2005) *Experiencescapes: Tourism, Culture, and Economy* (Copenhagen: Copenhagen Business School Press).

Pine, B. J. & Gilmore, J. H. (1999) *The Experience Economy: Work Is Theatre and Every Business a Stage* (Cambridge, MA: Harvard Business School Press).

Pratt, A. C. (1953–1974) The cultural industries production system: A case study of the employment change in Britain, *Environment and Planning A*, 29(11), pp. 1953–1974.

Pratt, A. C. (2005) Cultural industries and public policy, *International Journal of Cultural Policy*, 11(1), pp. 31–44.

Regeringen (2003) *Danmark i kultur- og oplevelsesøkonomien—5 nye skridt på vejen* (Copenhagen: Schultz).

Richards, G. (2001) *Cultural Attractions and European Tourism* (Wallingford: CABI International).

Roberts, K. (2004) *The Leisure Industries* (Haywood: Palgrave Macmillan).

Roberts, L. & Hall, D. (2004) Consuming the countryside: Marketing for "rural tourism", *Journal of Vacation Marketing*, 10(3), pp. 253–263.

Scott, A. J. & Power, D. (Eds) (2004) *Cultural Industries and the Production of Culture* (London: Routledge).

Smidt-Jensen, S. (2008) Kreative mellemstore byer, *Byplan*, 59(1), pp. 168–171.

Sørensen, F., Sundbo, J., Fuglsang, L., Svabo, C. & Darmer, P. (2007) *Udvikling i provinsbyer baseret på oplevelsesøkonomi og den kreative klasse* (Roskilde/Nykøbing F: Center for oplevelsesforskning/Centre for Leisure Management).

Sundbo, J. & Bærenholdt, J. O. (2007) *Indledning: Den mangfoldige oplevelsesøkonomi*, in: J. O. Sundbo & J. Bærenholdt (Eds) *Oplevelsesøkonomi: Produktion forbrug kultur* (Copenhagen: Samfundslitteratur).

Therkildsen, H. P., Hansen, C. J. & Lorentzen, A. (2009) The experience economy and the transformation of urban governance and planning, *European Planning Studies*, 17(6), pp. 925–941.

Thrift, N. (2005) *Knowing Capitalism* (London: Sage).

Vaekstfonden (2007) *Oplevelsesindustrien — perspektiver for iværksætteri og venturekapital* (København: Vækstfonden).

Weiermair, K. & Mathies, C. (Eds) (2004) *The Tourism and Leisure Industry: Shaping the Future* (Binghamton, NY: Haworth).

Winther, L. & Hansen, H. K. (2006) The economic geographies of the outer city: Industrial dynamics and imaginary spaces of location in Copenhagen, *European Planning Studies*, 14(10), pp. 1387–1406.

Appendix 1. Employment in Denmark, absolute numbers (selected years)

Employment, all branches	1993	2002	2006
Small municipalities, <20,000 inhabitants	869,851	905,719	891,452
Medium-sized municipalities, 20,000–100,000 inhabitants	1,048,913	1,119,249	1,118,433
Large municipalities, <100,000 inhabitants	679,610	745,942	736,890
Denmark	2,598,374	2,770,910	2,746,775

Source: Statistics Denmark and own calculations.

Appendix 2. Employment in hotels and restaurants and entertainment, culture and sports in Denmark, absolute numbers (selected years)

	1993	Share[a]	2002	Share	2006	Share
Hotels and restaurants (NACE 55)						
Small municipalities, <20,000 inhabitants	23,447	2.7	24,388	2.8	25,585	2.9
Medium-sized municipalities, 20,000–100,000 inhabitants	25,175	2.4	27,419	2.6	30,382	2.8
Large municipalities, <100,000 inhabitants	19,378	2.9	21,057	3.1	29,602	4.1
Denmark	68,000	2.6	72,864	2.8	85,569	3.2
Entertainment, culture and sports (NACE 92)						
Small municipalities, <20,000 inhabitants	8407	1.0	9221	1.0	11,975	1.4
Medium-sized municipalities, 20,000–100,000 inhabitants	15,766	1.5	17,117	1.6	21,641	2.0
Large municipalities, <100,000 inhabitants	17,453	2.6	18,018	2.6	23,537	3.3
Denmark	41,626	1.6	44,356	1.7	57,153	2.1
AEPS (NACE 55+NACE 92)						
Small municipalities, <20,000 inhabitants	31,854	3.7	33,609	3.8	37,560	4.3
Medium-sized municipalities, 20,000–100,000 inhabitants	40,941	3.9	44,536	4.2	52,023	4.7
Large municipalities, <100,000 inhabitants	36,831	5.4	39,075	5.7	53,139	7.3
Denmark	109,626	4.2	117,220	4.4	142,722	5.3

Source: Statistics Denmark and own calculations.

[a]Share of employment in selected branches of total employment in the indicated year and category of municipalities.

Appendix 3. Level of education for persons employed in hotels and restaurants, entertainment, culture and sports and all branches, in Denmark, absolute numbers (2006)

	Group 1	Group 2	Group 3	Not specified	Total
Hotels and restaurants					
Small	4910	8993	1580	534	16,017
Medium	6024	12,613	2026	555	21,218
Large	4557	14,225	4063	692	23,537
Denmark	15,491	35,831	7669	1781	60,772
Entertainment, culture and sports					
Small	14,467	13,200	885	1505	30,057
Medium	13,678	13,496	521	1487	29,182
Large	9926	16,091	1444	2141	29,602
Denmark	38,071	42,787	2850	5133	88,841
	Group 1	Group 2	Group 3	Group 4	Total
All branches					
Small	271,111	553,336	33,579	21,765	879,791
Medium	273,329	734,260	87,211	23,633	1,118,433
Large	137,687	466,251	114,335	18,617	736,890
Denmark	682,127	1,753,847	235,125	64,015	2,735,114

Source: Statistics Denmark and own calculations.

The Experience City: Planning of Hybrid Cultural Projects

GITTE MARLING, OLE B. JENSEN & HANS KIIB

ABSTRACT   *This article takes its point of departure in the pressure of the experience economy on European cities—a pressure which in recent years has found its expression in a number of comprehensive transformations of the physical and architectural environments, and new eventscapes related to fun and cultural experience are emerging. The physical, cultural and democratic consequences of this development are discussed in the article, as well as the problems and the new opportunities in the "experience city". It focuses on the design of the "Danish experience city" with a special emphasis on hybrid cultural projects and on performative urban spaces. "Hybrid cultural projects" are characterized by a conscious fusion between urban transformation and new knowledge centres, cultural institutions and experience environments. "Performative urban spaces" are characterized by stages for performance, learning and experience. The performative activity can take on the guise of events—something temporary, but still recurring, which influences the shape and identity of the space. The article contains four sections. In the first section, we present three European cases outside Denmark in order to relate to the wider international debate and development. In Section 2, we present the main theoretical concepts and framings that will guide the understanding and the analysis of the experience city. In Section 3, we focus on the design of the "Danish experience city" and present the first research findings. The projects are categorized according to their content, structure and urban localization. In particular, the cases are labelled in relation to their strategic and urban planning importance, their social and cultural content and their architectural representation and the programmes they contain. The article ends in a discussion of the trajectory for future research.*

1. Introduction

Many cities are currently supplementing the central shopping areas with new event spaces, pedestrian streets and café environments. Experience and entertainment have become main priorities. But it is not only the pedestrian streets and the classic locations in the old city centres that are affected by the modern city dwellers' demand for experiences. The transformation of the old industrial areas, the water fronts and the recreational facilities in the vicinity of the cities are also a part of this demand. The buzzword in this context is the development of an "experience economy" as the driving force behind these transformations (Kiib, 2007). Strategic planning efforts are focused on the experience economy as an essential factor in developing the physical and cultural image of the city as well as its economic basis (Hall, 2000; Landry, 2000; Evans, 2001; Metz, 2002; Boer & Dijkstra, 2003; Kunz-mann, 2004; Zerlang, 2004).

The current societal transformation processes of globalization, increased regional and global commerce and mobility patterns and a general shift towards the immaterial side of production are all vital background dimensions to the "experience city". Harvey speaks of a transformation from Fordism to "flexible accumulation" (Harvey, 1990, p. 41) as a way of understanding the transformations of the organizational and spatial frame of capitalism. However, the way capital works in its changed relationship between the material and the immaterial has also called for new terminologies, as, for example, when Lash and Urry (1994, p. 60) identify a new regime of "reflexive accumu-lation". Diverse theorists have argued for a renewed understanding of the "knowledge-based economy" (Jessop, 2004, p. 49), and the "informational mode" of capitalism (Castells, 1996) that together creates "Experiencescapes" illustrative of the material and immaterial dimension of the new economy (O'Dell & Billing, 2005). In this article, "experience" covers many analogous concepts such as discovery and practice, to live through something. And as a consequence of the experience, one will be skilled, experi-enced, competent and tested. In other words, there is an element of learning, refinement and culture which is often ignored in the more marketing-oriented discourse of the experi-ence economy and experience city. In this sense, the "experience economy" places new demands on urban political strategy-making, on the cities' cultural networks, on local artistic competences and on the spatial and architectural frames (Kiib, 2006). The physical, architectural and planning-oriented dimensions of flexible accumulation, reflexive accumulation and experiencescapes are captured by the notion of "hybrid cultural project" and "performative urban space". We define "hybrid cultural projects" as a con-scious fusion between urban transformation and new knowledge centres, cultural insti-tutions and experience environments. "Performative urban spaces" are defined as stages for performance, learning and experience. The performative activity may be a temporary event but still recurring, which influences the shape and identity of the space.

To focus explicitly on Danish cases of experience city, design and policy-making is a deliberate choice making comparisons less evident, but on the other hand bringing to the foe the particularities of the Scandinavian Welfare City and how this is intimately related to the practices of experience city design. Moreover, we refer to international cases in this paper, but more as a precursor to the debate and empirical cases which are all Danish.

2. Culture and Experiences as Generators of Urban Development

In recent years, almost all European cities have been alert to the potentials within the new experience economy. In particular, how such a shift might fertilize urban transformation and development which is less tied to the industrial society and more to an experience-oriented economic development. In this section, we briefly present three examples that each in their own way has become a landmark example for many European cities and municipalities. Each of these three narratives from the European scene suggests that there is ongoing work taking place which focuses on hybrid cultural programmes in an attempt to re-develop cities and districts. The narratives bare witness to an understanding of the experience economy as something with great potentials when coupled with "strategic thinking and cooperation between public and private sector agencies". The cases testify that "architectural icons, urban renewal and performative urban spaces" are used as a tool for city branding.

2.1 *Emscher Park*

Emscher Park is an area near the Emscher River in the German Ruhr district (Figure 1). It is an example of how innovative city planning and urban design and the experience economy have lead to an entirely new agenda for the area. For decades, the area has been one of Europe's largest industrial areas with mining operations and heavy industry. It has been notorious for severe environmental problems due to fumes, noise and air pollution—and it has been infamous for its problematic social conditions and high unemployment rates. The area has commenced a long process of transformation in which the old industrial architecture is preserved and plays a decisive part as a framework for new experiences and new forms of cultural tourism. Presently, these industrial plants emerge as rusty dinosaurs in the middle of green recreational facilities. Some appear only as landmarks for the area, as sculptures in which one can enter and walk or admire as lightshows with their spectacular appearance accentuated during the night. Other buildings have been converted into exhibition facilities, show rooms, workshops for artistic production, casinos, theatres and facilities for sports and education (Braa, 2003).

The Emscher Park project has created a new "brand" and given rise to a new pride in an area which has had a very bad reputation—despite the fact that it was here that the economic basis for the West German welfare society was created after the Second World War. By

Figure 1. Emscher Park: architecture without architects (author's photo)

fusing the logic of the experience economy with principles of environmental sustainability, a platform for the new development of the area has been successfully created. The Emscher Park project illustrates that the experience city and sustainability are not necessarily mutually exclusive strategies (Marling & Kiib, 2007). As such, Emscher Park is illustrative of our interest in de-constructing the taken-for-granted understandings of the experience economy.

2.2 *Barcelona*

Barcelona is noteworthy due to its conscious effort during the last 15–20 years to use the experience economy as the driving force in transformation strategies. The city has had to face huge demands in terms of accessibility planning, but the demands for aesthetic and architectural quality have also been considerable. The city's planners and decision-makers have realized that in order to make an event attractive or invite people to stay in the urban spaces, it is necessary to create welcoming surroundings that appeal to a variety of social groups and which are functionally suited to a variety of applications. Throughout the entire period, Barcelona worked with an urban space strategy that has transformed the massive building structure of the city into a web of spaces, small and large, that sets the stage for a public and democratic urban life which was unheard of during General Franco's regime (Figure 2). This has been achieved by transforming local industrial areas into new parks and squares or by demolishing selected housing blocks in order to let in light and air and create urban "breathing holes" (Marling & Kiib, 2007).

Barcelona has been aiming high. The visions were given full scope in connection with the grandiose plans to organize the 1992 Olympics where traffic installations and the Olympic village at the new waterfront contributed to the establishment of the new structural transformation of the city. The city goes on branding itself as one of the new strong

Figure 2. Barcelona forum by Herzog and de Meuron (author's photo)

European cultural cities. This is achieved with the aid of a variety of tools, such as the urban space strategy, but also strategies related to the construction of new museums, exhibition facilities and forums for architecture and art. Additionally, this is followed-up by a number of innovative cultural programmes and events (Gehl & Gemzøe, 2000; Marling & Kiib, 2007). In Barcelona, the development has enabled a shift in economic structure towards more emphasis on the experience economy. Importantly, however, the development has also sent Barcelona on its way back towards being a democratic urban community with a pluralistic and public urban life and a new cultural elite. Thus, city centre design, urban space strategies, waterfront transformations and other similar projects are currently part of planning policies all over Europe (Gehl & Gemzøe, 2000).

2.3 *Graz*

The cultural capital usually also focuses on new architecturally ground-breaking buildings, as, for example, in Graz where the new experience bridge offers the citizens a space to stay, play, perform and have a cup of coffee in the middle of the city's flowing river. Architect Peter Cook's Kunsthalle, which is wedged in like a foreign element with interactive facades, is a provocative architectural artefact which has become a new symbol of the city (Figure 3). It couples experience with the physical urban environment in a new and interesting manner—a form of interaction that is bound to become a much more familiar sight (Kendahl, 2003).

In Graz, a number of new venues offering the opportunity to experience the city and the culture from new angles have been created. However, the new buildings have left the city with a large economic burden in the form of operating expenses that drain the municipal coffers and will prevent the city from funding other types of cultural initiatives for many years to come. The example also shows that it is relevant to ask what will become of the social and cultural network that has been established during the culture capital year. Is

Figure 3. Kusnthalle Graz by Peter Cook (author's photo)

it possible to maintain and expand it? Or has it all ended in conflict related to financial problems?

2.4 *Learning and Research Questions*

The learning from these international cases and the success of the implementation of the strategies behind them raise a number of research questions relevant to this project. We argue that the projects can be labelled as "hybrid cultural project" with three clear dimensions: (1) a new "urban political agenda and strategy for territorial re-development"; (2) a new "strategy for urban cultural life" and finally (3) the development of new "architectural typologies and urban spaces".

2.4.1 *A new urban political agenda and strategy for territorial re-development*
All these projects contribute to a "new urban political agenda" within the current debate about the future of our cities in "the experience economy". A strategic urban transition from industry to knowledge, sport and leisure seems to be a keystone in this strategy. In all three international cases, a number of "lighthouse projects" have been developed. But as we see in Barcelona, these have been linked to "a linear re-development strategy" related to infrastructural development and along the coastline, where the Olympics Village, a number of public parks, new knowledge institutions, leisure capes and housing projects have been developed. In Emscher Park, the strategic thinking has been linked to a "field-strategy" with a re-development related to "nodes" and "voids" in a city network. This learning raises research questions with respect to how the new urban political agenda is linked to different territorial strategies, e.g. a "lighthouse strategy", a "linear strategy" or strategies related to "nodes" or "voids", in a city network?

2.4.2 *New strategies for urban cultural life*
The new projects have become the hotbed of a new urban culture that consciously fuses the traditional shopping and café life of the city with knowledge, experience and play. In the Cultural City Project 2003 in Graz, the focus has been on the development and attraction of the cultural elite and the artistic avant-garde. The city has promoted itself as a high-ranked hub for music and performative art. But other cities such as Barcelona have had a more comprehensive social strategy—also related to the cultural public. Leisure and tourism have been linked with hundreds of new public squares meant for community development all over Barcelona, and recently this has also been linked to new concepts for the development of urban environments for learning—"the knowledge city". In Emscher Park, the industrial heritage has been very much in focus, and the linkage between learning from the past and the future life of the local dwellers has been the platform for new developments. Such development raises a number of research questions on how the different strategies for urban cultural life are linked to different cultural groups and lifestyles—including the "cultural elite", the "artistic avant-garde", the "cultural public" or "subcultures"?

2.4.3 *Development of new architectural typologies and urban spaces*
We have found that the projects contribute to the development of "new architectural typologies and urban spaces". They often draw upon the scale and typological multiplicity of the industrial architecture and harbour environments, and through a new orchestration, they both enrich our architectural heritage and create new architectural projects focusing

on dialogue and transparency between old and new architectural typologies. Kunsthaus Graz is an "iconic" project with a new type of "performative architecture". The architectural language is not only challenging and contrasting the nineteenth-century typology of the site, in itself it "performs" and "speaks to you", and in an inclusive way, it invites people into new architectural spaces enhancing bodily and artistic experiences. This kind of "new architectural performance" is also present in a number of urban space projects in Barcelona. In a way, the projects are on track in the development of a new architectural typology with interactive facades and with performative "intelligence".

Not all projects are related to new architectural form. Actually many projects are developed in abandoned large-scale industrial complexes. In this way, many "iconic" projects gain life from the "architectural narratives of industrial heritage". Other projects again are more related to events and festivals. These projects use concepts for temporary architecture or a new form of "instant urbanism". This raises a number of research questions on how the different strategies for urban cultural life is enhanced by different architectural representations and narratives—including "new iconic architecture", "architectural heritage", "performance architecture" or "temporary architecture without architects".

As stated in Section 1, the international cases lead us to a thesis that the strength of "hybrid cultural projects" is the conscious combination between learning and playing, between public and private and between artistic quality and the popular activity. The starting point is a common willingness to include many different groups, and at the centre of it all, a dynamic hybrid of "edutainment", high culture and bodily exertion is challenging our traditional perception of urban life. It is a thesis that "performative urban spaces" in conjunction with the hybrid cultural projects can latch on to a variety of different purposes that have to do with cultural understanding and exchange. In these projects, we find a certain drive towards a public urban life that does not merely encompass the well-off and the well-educated parts of the population; some of the projects have the potential to include and activate newcomers to the society, the young ones, the old ones, the not so well-adapted, etc. Thus, it is the overall thesis that these hybrid cultural projects, stages and spaces are potential "public domains", i.e. places for social and cultural exchange between lifestyle groups with different values and worldviews (Hajer & Reijndorp, 2001).

3. Framing the "Experience City"

In the theoretical framing of the "experience city", we will focus on two areas which we believe would be of particular interest seen from a planning perspective. First, we discuss "planning and design within the experience city". Under this theme, the issues are which role hybrid cultural projects can play in the urban transformation process and how they impact public planning practice agencies and processes. This theme also has to do with how the projects influence the urban architecture and the physical planning of the city. The second area has to do with "the social and cultural programmes of the experience city". Under this theme, issues of social inclusion and cultural diversity are at the centre. In the following section, we present key ideas within such a framework.

3.1 Planning and Design Within the Experience City

In the midst of this context of change, cities and urban policy makes have come to re-orient the tasks of urban governance and planning in the direction of "luring potential capital into the area" (Rogerson, 1999, p. 971). The mainstream assumption is that within this field of increased inter-urban competition, leadership and risk-taking attitudes are the keys to success (Lever, 1999, p. 1042). Moreover, the assumption seemed to be that the less tangible dimensions of urban life were not just everyday life conditions to urban dwellers, but also vital competitive parameters:

> As interurban competition on a global scale became the norm in the 1980's and 1990's, image took on an ever more vital role in urban economies ... "Quality of life" became the rallying cry of many big-city mayors elected at this time, based on a "broken window" theory whereby the simple appearance of disorder had a material effect of provoking criminal behaviour, thus justifying urban policies based more on cleaning up those appearances than on addressing underlying social issues. (Greenberg, 2000, p. 250)

The number of strategic and proactive "responses" to the situation of increased global urban competition has given rise to the notion of the "entrepreneurial city" (Hall & Hubbard, 1998). Jessop explores this intricate but important relationship between the material base of urban economy and its symbolic forms of representations:

> ... The city is being re-imagined—or re-imaged—as an economic, political and cultural entity which must seek to undertake entrepreneurial activities to enhance its competitiveness; and ... this re-imag(in)ing is closely linked to the re-design of governance mechanisms involving the city—especially through new forms of public-private partnership and networks. This is evident in the wide range of self-representational material emitted by cities and/or agencies involved in their governance. (Jessop, 1997, p. 40)

The contemporary urban situation is thus marked by fundamental transformations in the cultural, economic and social basis for urban life. The question is how this relates to the making of urban spaces and architecture. One of the most direct attempts to deal with this question is the work of Klingmann. According to Klingmann (2007, p. 19), "designing for experience requires connecting architecture to the user's personal dreams and desires". Accordingly, architecture and urban interventions may contribute to engaging its public on the level of the senses in meaningful connections (Klingmann, 2007, p. 51) which move beyond the iconic buildings so characteristic of much contemporary city politics (Jencks, 2005). The city may be discussed within the framework of Klingmann's as a site of use, symbolism and experience. Put differently, we may ask how city design and urban spaces produces use value, symbolic value or experience value? (Klingmann, 2007, p. 6). This means that urban spaces and interventions herein may not only have an important use value and symbolic dimension, but also an (perhaps under-theorized) experience value (Klingmann, 2007, p. 44).

With reference to Bernd Schmitt's analysis of marketing, we may want to re-phrase our understanding of architecture (and urban intervention) in the experience economy along four distinct dimensions leading to a fifth level of synthesis. In Klingmann's terminology, we would be looking at a synthesis of "sense architecture", "feel architecture", "think architecture" and "act architecture". Assuming that these four dimensions all merges into one synthesis, we may start to speak of "relate architecture" in which sensing (our bodily engagement with spaces), feeling (our emotional engagement with spaces), thinking (our cognitive and reflective engagement with spaces) and acting (our socially transformative engagement with spaces) come together in a strong emotional and cognitive relation between subjects and urban architectures (Klingmann, 2007, p. 50). In the experience economy, urban interventions may thus facilitate new deliberation processes and forms of interaction that point towards progressive experiments and hybrid socializations. Put differently:

> When applied to architecture, Schmitt's modules imply that in order to create an architecture that bonds with people in their daily lives, attention needs to be refocused on the transformative dimension of space and the emotions created by its use (Klingmann, 2007, pp. 50–51)

However, the analysis of the experience city must also include a more critical perspective acknowledging that these urban interventions run the risk of fuelling social exclusion, cultural homogeneity and a culture of fear of the other (Marling & Zerlang, 2007, pp. 6–7). Next to the new exciting possibilities of experience interventions (Marling, 2003), there are critical issues concerning the democratically legitimate base for branding, notions of identity production and the protection of minorities, as well as issues of social inclusion and the commodification of the city (Jensen, 2007b, p. 118). Kvorning places this critical dimension on the new approach of cultural planning centrally as he poses the question:

> Are there other ways of dealing with these questions, the fear of the stranger, are there any other ways than the Disney way to deal re-establishing the system of learning from the stranger, are there ways of creating zones which can start a new process? That must be the key question for cultural planning. (Kvorning, 2004, p. 55)

The creation of new urban interventions aiming at satisfying the demand for experience and stimulus does not necessarily need to be commercial and instrumental. Rather, the more critical insights of contemporary urban theory point at the potential for creating learning environments and situations where multiple and heterogeneous social groups may create new public domains. The notion of "public domain" is here understood as: "places where exchange between different social groups is possible and also actually occurs" (Hajer & Reijndorp, 2001, p. 11). Thus, hybrid cultural projects and performative urban spaces may be thought of as sites of "learning from the stranger" and places of civil society based interaction (regardless of the fact that numerous projects stay firmly on the side of economic opportunity). In the words of Hajer and Reijndorp (2001, p. 88): "public domain is thus not so much a place as an experience". In the "post-spectacular city", what matters is no longer architecture as spectacle and icon, but architecture as activity (Thackara, 2005, p. 185).

Summing up on the theme of planning and design within the experience city, we would point to a need for a supplement to the traditional "master plans". This means developing planning tools which are more open, dynamic and which cuts across sectors (Hall & Hubbard, 1998; Lever, 1999). Related to this, there is a demand for organizations capable of handling such a wide arena of topics and agencies. Moreover, there is a need to break away from the established ways of thinking by "stirring the city"

(Bunchoten, 2001) in order to re-connect agencies and agents of the city in novel and creative ways. Lastly, we see a demand for an open and engaging type of urban design and
architecture which partly reflects the hybrid programmes and partly the diverse target
groups present in the contemporary city (Hajer & Reijndorp, 2001; Klingmann, 2007).

3.2 *Social and Cultural Programmes of the Experience City*

The second theme we want to address in the theoretical framing is the notion of social and
cultural programmes in the experience city. Here we get closer to the psychological and
experiential dimensions of the new economy. Thinking about the social and cultural programmes in the city means to understand how the new experience economy is constituted
and what this might mean to people.

The German sociologist Gerhard Schulze pointed at the new orientation towards experiences and the social pragmatism related hereto in the early 1990s. To Schulze, the prime
factor in the transformation from the industrial society governed by tradition to a new way
of engaging with notions of belonging, identity and everyday life is the individual's
experience of instant meaningfulness and stimulation: "Erlebnisorientierung ist die unmit-
telbarste Form der Suche nach Glück" (Schulze, 1992, p. 14). In a post-scarcity society
(Schulze do acknowledge that also within rich Western societies there are vast inequal-
ities), the mental and practical orientation of social individuals is increasingly being
defined by fun, exciting and stimulating activities and experiences. Balancing between
risks of boredom and insecurity, the individual of the experience society seeks stimulation.
This orientation towards experience seen as act of individual stimulation relates to the way
urban spaces and architecture is less about the formal properties of the "object" and more
about the effects it generates for the "subject" (Klingmann, 2007, p. 11). The preoccupa-
tion with semiotics, signs and branding therefore, connects to the ephemeral and liquid
dimension of contemporary capitalism:

> The societal transformation process of Western countries has been characterised by a
> shift towards immaterial and experiential stimulation. Even though there are massive
> inequalities and welfare problems, the global shift has given completely new tools to
> social agents, both for constructing identities and relating to one another … As
> experience and culture gain importance, cities world wide are engaged in construct-
> ing images and representations of their locations in accordance with these new trends.
> Therefore the culture-led, experience-oriented policy makers are looking towards the
> discipline of urban branding. (Jensen, 2007a, p. 212)

In the literature, the much quoted notion of "experience economy" obviously is a central
layer next to the existentialist notion found in Schulze's work. To Pine and Gilmore (1999,
pp. 6–14) the hallmark of our economy is that it is an "experience economy". Accordingly, there is a very high added value in moving from commodities, goods and services
into the "fourth dimension", namely that of the "experience". Adding the symbolic dimension of a nice café atmosphere makes the ordinary cup of coffee multiply the revenue
potential (Pine & Gilmore, 1999). Stressing the open-endedness of their way of seeing
the new economy, they boldly state that "there is no such thing as an artificial experience"
(Pine & Gilmore, 1999, p. 37), which is a clear suggestion that the sky is the limit in terms
of turning the city into ever-new sites to be consumed.

But there is more to the ephemeral and immaterial dimension of the experience economy than adding atmosphere and themed environments. The presence of social actors, creative groups and institutions seems just as important. In the words of Landry, there must be a "creative milieu" which is:

> ... a place—either a cluster of buildings, a part of a city, a city as a whole or a region—that contains the necessary preconditions in terms of "hard" and "soft" infrastructure to generate a flow of ideas and inventions. Such a milieu is a physical setting where a critical mass of entrepreneurs, intellectuals social activists, artists, administrators, power brokers or students can operate in an open-minded, cosmopolitan context and where face to face interaction creates new ideas, artefacts, products, services and institutions and as a consequence contributes to economic success. (Landry, 2000, p. 133)

And moreover, the clusters of creative agents are seconded by an intense seeking of fun and ludic experiences by the urban dweller as it becomes clear that leisure is not a secondary activity of cities in the new economy:

> Leisure is more than the time you can spend as you like, it has become an omnipresent culture of fun with an enormous economic importance. Our social identity is determined by the way we spend our leisure at least as much as by the work we do or the possessions we own. (Metz, 2002, p. 8)

Without going deeply into this, one of the key urban leisure activities of the contemporary city must be acknowledged to be shopping or even the merging of entertainment and shopping into "fun-shopping" (Boer & Dijkstra, 2003, p. 185).

From this discussion, and related to the previous, it seems to us that the democratic "Welfare City" needs to strengthen its diversity. This might mean overcoming the fear of the "strangers" by creating common experiences and facilitating curiosity and mutual learning (Bauman, 2002). Indeed, there is a need for public domains to facilitate this, and there is a need to develop new dynamic cultural prototypes that are socially inclusive (Marling & Kiib, 2007). But it also becomes evident that events and cultural programmes which can absorb the flux and dynamics of the experience city are of great importance. Often this must be done by means of strategies for nurturing creative groups and artists (Landry, 2000).

Summing up on this theoretical framing, we would argue that one should pay attention to a number of dimensions in the empirical analysis of the cases. First, this would mean looking into questions of how the institutional make-up in general is configured, including the interplay between the new urban agenda and the promoted urban re-development strategies—what are the prevailing rationales underpinning the projects and interventions. Secondly, we would be interested in showing how the interventions facilitate new cultural practices. We should enquire about the kinds of interaction and experiences enabled (or constrained) by these urban interventions as well as the sort of interactions and mutual learning processes that will be shaped in and by the projects. Thirdly, we are in search of new symbolic interpretations of these projects and especially the impact on the development of urban spaces and new performative architecture. These are the underlying dimensions of the cases to be explored that grow out of the theoretical framing.

In the exploration of the Danish projects to follow, we will primarily be focusing on the project's strategic role in urban transformation and planning. Moreover, we will focus on the role of the projects in the urban architectural context and the importance of the cultural programmes in building a diverse cultural life and new public domains. From the theoretical framing of the experience city and its empirical expression in the selected European examples, we now turn towards the exploration of Danish cases of experience design.

4. Designing the Danish Experience City

Also in Denmark, we are currently witnessing the appearance of many new projects related to culture, sports and art. In the light of our definition of the "experience city" in the introduction to this article, there are many projects that seem particularly interesting.

4.1 *Sixteen Danish Projects*

The research project exploring the design of the experience city is based upon quantitative and qualitative research conducted in autumn 2007 and spring 2008. The quantitative part contains a survey sent to all 98 Danish municipalities in October 2007. The questionnaire was distributed by e-mail and contained four main themes: cultural projects, urban spaces, urban regenerations and cultural events. The municipalities were asked to underline which projects they found particularly important as expressions of the experience economy. The surveys were in general answered by civil servants in the technical and cultural departments of the municipalities. Sixty-one municipalities responded, which equals a response rate of 62%. After the quantitative enquiry, a number of qualitative research interviews were conducted in the spring of 2008. The themes and issues are the ones structuring the case-overview table in this article (Table 1). In general, we interviewed municipal planning officers and people involved in the genesis of the projects (both planners and grass-root activists). The following analysis is based on the quantitative results of the general overview and the qualitative results for the specific details concerning each of the 16 projects.

The survey and interviews testify that municipalities are directly engaged in cultural and experience strategy-making and planning. In the overview of the projects, it has been interesting to note that all cities and municipalities have engaged with the experience economy and in a diverse set of ways attempted to make it a dynamo for city branding, urban transformation and cultural development. Many interesting small and larger projects have thus seen the light of day. Out of these, we have selected 16 larger projects with the main characteristics that we find of particular interest in relation to the research questions. The 16 projects are listed with their main characteristics described (Table 1). The projects are divided into three overall categories: buildings and urban architecture, performative urban spaces and temporary cities/urban events.

As we are engaged in understanding the ways that the projects play different roles in the urban transformation and strategic planning process, we are approaching this from a research strategy of labelling or "naming" some of the main characteristics. Needless to say, this is not a completely finished process and the coining of concepts and labels may most certainly be discussed. The fact that they are first interpretations of the empirical material is a fundamental feature of the research and thus invites other interpretations as well. In this phase of the research into designs of the experience city, we see it as a vital

Table 1. Selected projects of Danish experience city design

Projects	Role in planning and urban development	Cultural programmes and target groups	Architecture
1. Nordkraft, Aalborg Transformation of a former power plant at the water front	Urban strategy: lighthouse project	Target group: cultural public/cultural elite	Architectural status: icon/historic monument
Rhythmic music, cinemas, media, sports, leisure, shopping	Agents: public (many agents)	Programmes: many and diverse	Interval: permanent
2. Paper factory, Silkeborg Transformation of a former factory plant at the stream	Urban strategy: lighthouse project	Target group: cultural elite	Architectural status: historic monument
Culture, cinema, institutions, business, hotel, housing, fitness and workshops	Agents: private/public (many agents)	Programmes: many and diverse	Interval: permanent
3. Alsion, Sønderborg New architecture at the stream	Urban strategy: lighthouse project	Target group: cultural elite	Architectural status: icon (modernistic architecture)
University, train terminal, science park, concert hall, restaurants	Agents: public (many agents)	Programmes: many and diverse	Interval: permanent
4. Music theater, Holstebro Redesigned architecture in the city centre	Urban strategy: lighthouse project	Target group: cultural public	Architectural status: container (a cluster of buildings with different architecture)
Performance, music, dance, education	Agents: public–private partnership (many agents)	Programmes: many and diverse	Interval: permanent
5. Bazar Fyn, Odense Transformation of a factory site in the core of the city	Urban strategy: linear attractor (new development)	Target group: subculture	Architectural status: container (historic buildings)
Bazar, exotic market, café's, business incubators	Agents: private partnership (many agents)	Programmes: many and homogenous (related to business, leisure and shopping)	Interval: permanent

(*Continued*)

Table 1. Continued

Projects	Role in planning and urban development	Cultural programmes and target groups	Architecture
6. Nicolai, Kolding Collection of public cultural institutions and housing in old school close to city centre: cinema, cultural house for children, café, concerts, exhibition spaces	Urban strategy: lighthouse Agents: public partnership (many agents)	Target group: cultural elite/cultural public Programmes: many and homogeneous (related to public service and culture)	Architectural status: container (historic buildings) Interval: permanent
7. Brandts, Odense Three institutions (museums+art exhibition)+learning labarotory and shop. Events	Urban strategy: lighthouse Agents: public partnership (many agents)	Target group: cultural elite/cultural public Programmes: many and homogeneous (related to exhibitions)	Architectural status: container/icon (historic buildings) Interval: permanent
8. Islands Brygge, Copenhagen People's park at a brown field in city harbour Free swimming, skating, barbecue, beaches and culture house	Urban strategy: field/void Agents: private–public partnership	Target group: cultural public/subcultures Programmes: many (related to public service and culture)	Architectural status: icon (historic ruins) Interval: permanent (activities by seasons)
9. Solbjeg Square, Copenhagen A former railway area transformed into a square with performative elements as light, steam and green elements	Urban strategy: void Agents: public	Target group: cultural public Programmes: few (related to transit and park)	Architectural status: icon Interval: permanent (activities by seasons)
10. Underværket, Randers A covered square and part of a block transformed into a cultural meeting place with shops, cafe's education and events	Urban strategy: lighthouse (small) Agents: public	Target group: subculture/cultural public Programmes: few and diverse (related to shopping, events and public service)	Architectural status: container Interval: permanent

11. Amagerbrogade, Copenhagen	Urban strategy: linear attractor	Target group: cultural avant-garde/ cultural public	Architectural status: icon (interactive spaces)
New performative places along the street, story-telling. Multiple programmes Combining different neighbourhoods to the street			
Interactive technologies	Agents: public (collaboration with local citizens)	Programmes: few and diverse (related to shopping, events and public service)	Interval: permanent
12. Roskilde Festival, Roskilde	Urban strategy: field (planning by coincidence)	Target group: cultural avant-garde/ subculture (Jung people)	Architectural status: temporary architecture
Music festival with 80,000 participants Music venues, cinema, theatre, events, exhibitions			
	Agents: private (many NGO agents)	Programmes: many and homogeneous (related to rhythmic music and culture)	Interval: temporal (1 month city)
13. Festival week, Aarhus	Urban strategy: field (cultural strategy and city branding)	Target group: cultural avant-garde/ subculture (Jung people)	Architectural status: temporary architecture
One week open and inviting event in the city			
Exposes culture as part of urban transition with theatre, cultural events, music, street art and performance	Agents: public–private partnership (many NGO agents)	Programmes: many and homogeneous (related to rhythmic music and culture)	Interval: temporal (1 week city)
14. Grey Hall & Loppen, Christiania	Urban strategy: void /field	Target group: subculture/cultural public	Architectural status: stated as temporary about 30 years ago
Music, performance, art exhibition, events (Christmas eve for homeless people), market, meeting place			
	Agents: private partnership (many NGO agents)	Programmes: many and homogeneous (related to rhythmic music, culture and social events)	Interval: permanent

(Continued)

Table 1. Continued

Projects	Role in planning and urban development	Cultural programmes and target groups	Architecture
15. Metropolis, Copenhagen			
Ten years programme of a series of cultural biennales with seminars, workshops, events, mobile sections around the city	Urban strategy: field (cultural strategy for performative art and experiments)	Target group: cultural avant-garde	Architectural status: temporary architecture
	Agents: private (many NGO agents)	Programmes: many and homogeneous (related to rhythmic music and culture)	Interval: temporal
16. Middle Age Days, Horsens			
Festival once year. Theme event circulating in the city. Attract 200,000 people from Denmark and Europe	Urban strategy: field (cultural branding strategy)	Target group: cultural public	Architectural status: temporary architecture
	Agents: private–public partnership (many NGO agents)	Programmes: many and homogeneous (related to shopping and culture)	Interval: temporal (once a year)

part of the research outcome to conceptualize and contribute to an adequate vocabulary that may help facilitating further analysis and research as well as public debate about the nature of these urban interventions.

The labelling will fall into three main categories. First, there is the "role of the projects in urban transformation and planning". Secondly, there is the "role of the projects as creator of new forms of culture, meeting places and public domains". And thirdly, we find the "relation of the projects to urban transformation and architecture" of importance. Importantly, this labelling is a heuristic devise as the real-life complexity of the cases may transcend the diversification of concepts, the implication being that some projects may be within more than one category.

4.2 *The Role of the Projects in Urban Transformation and Planning*

In relation to the role of the projects in urban transformation and planning, we find in the empirical material four main categories: lighthouse, linear attractor, field and void. We label projects "lighthouse" if the project contains a cluster of programmes localized within a relatively confined area in the city. These are expressions of a public strategic intervention and they are related to large investments. Lighthouse stirs large local attention and often has a massive local backing. They function as motors for existing activities and generate new cultural products and activities. Within the sample of projects presented in this article, Alsion in Sønderborg, The Paper Factory in Silkeborg and Nordkraft in Aalborg are illustrative examples of lighthouses. We coin the notion of "Linear attractor". By this is meant a series of widely different but related projects to be found along a line or a very long-stretched field. They express the goal of creating a new transformation and dynamics by connecting different elements (e.g. city neighbourhoods, programmes, etc.). As examples of linear attractors, Amagerbrogade and Solbjerg Plads in the Copenhagen area could be mentioned. The third category under the theme of urban transformation and planning is the "field". A field in this terminology identifies the cultural intervention spread over a larger area. It can refer to multiple touch-down locations in the city by different but related projects and events. A field can embed a broad form of intervention, but it can also be focused in time. Examples of fields are Aarhus Festival Week, Middle Age Days in Horsens and Metropolis in Copenhagen. The fourth and last category to be found under this theme is the "void". Voids are islands or sites of "negative force" within the urban fabric. A void is surrounded by fields attracting large investments and they represent a "hole" in the city's portfolio of strategies and interventions. By being a negation of the well-established and higher-ranked areas, the void attracts alternative social and cultural activities by coincidence. Simultaneously, the void functions as an incubator to alternative milieus and often works as site for exposed groups and the avant-garde. Often the void itself is locus for contestation and political manifestations. Examples of voids in the sample presented here are Christiania (Grey Hall & Loppen) and Roskilde Festivallen in Roskilde. Islands Brygge could also be seen as an example of a void with Peoples Park surrounded by an area with high investments.

To get an overview of the projects' relative positioning in relation to their role in the urban transformation and planning, we plotted them on to maps with the axes "lighthouse–void" and "private–public" (Figure 4). On the issues of strategic planning, we see that the projects tend to cluster at the "public–lighthouse" area and none in the opposite field of "private–void", suggesting (not surprisingly) a strong political lead in the

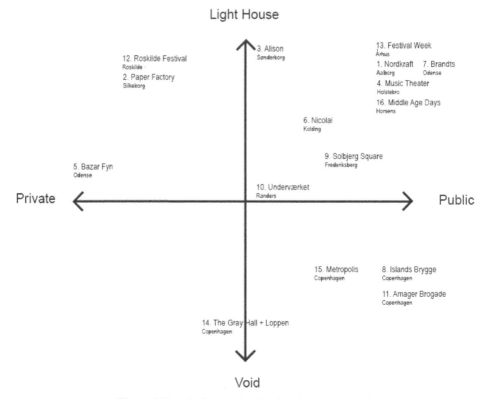

Figure 4. Level of strategic planning (at project start)

public projects. The diagram shows a fair distribution of projects in terms of private and public as well as lighthouse and void. This indicates that our case sample reflects the diversity we wish to investigate.

To explore the programmes and the target groups, we plot the projects in another diagram with the axes "few–many" and "elite–subculture" (Figure 5). In relation to the number of programmes planned, there is a large cluster of projects at the "elite–many" pole, suggesting that mainly the cultural elite and cultural public are considered and suggesting quite some variation in the types of projects (a diversity stemming from the multiple programmes).

4.3 *The Project's Role as Creator of New Forms of Culture, Meeting Places and Public Domains*

Under the theme of new forms of culture, meeting places and public domains, all the selected projects are aiming to gather more activities and programmes in order to establish a cultural synergy effect. Here we find the four categories of cultural elite, cultural avant-garde, cultural public and subcultures.

The polar category "elite–subculture" is depicted in Figure 5 where the categories are specifically exploring the relationship between programme and target group. The first one we would point our attention to is the "cultural elite". This is the nomenclature for projects

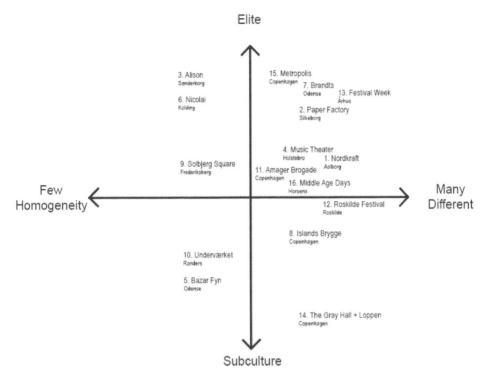

Figure 5. Programmes planned (at project start)

targeted at the middle classes, the well-educated and the economically well-off. The projects within this category seek to challenge the pre-understandings of the target groups, to stimulate to reflexion and afterthought, and they often contain elements of learning. Examples of cases within this category would be Alsion in Sønderborg and Papirfabrikken in Silkeborg. "Cultural avant-garde" describes the projects especially targeted at young people or the cultural avant-garde seeking challenges and "edge" in their cultural experiments. Examples are Metropolis in Copenhagen and Roskilde Festivallen. "Cultural public" labels broader projects in their audience appeal and aims at being socially inclusive by addressing themselves to different user groups. Examples are Nordkraft in Aalborg, Islands Brygge and Amagerbrogade in Copenhagen. And finally, the "subculture" category describes the projects that besides a general appeal have a special focus on social and sub-cultural groups and seek to facilitate a meeting with the broader population. Examples are Underværket in Randers, Grey Hall & Lopen (Christiania) in Copenhagen and Bazar Fyn in Odense.

4.4 *The Project's Relation to Urban Transformation and Architecture*

The relationship between physical transformation and architectural expression is the last theme we want to explore. Here we find the four categories of icon, historic monument, container and temporary architecture. "Icon" refers to those cultural projects that distinctly distinguish themselves from the architectonic context and which by its form constitutes a landmark (Lynch, 1960) or a focal point (Cullen, 1971). Generally, they have a large symbolic branding value to the city as a whole and strengthen the status and image of the local site of its placement. The "icon" may work as a driving force in the physical transformation of the city. Often icons are time-specific monuments and remarkable illustrations of contemporary architecture. Examples would be Alsion in Sønderborg, Solbjerg Plads in Frederiksberg and Islands Brygge in Copenhagen. The "historic monument" is the description of the architectural transformation of older listed buildings, for example, old industrial facilities being transformed to new cultural programmes. The historical monument carries the narrative of the city within and is often seen as positive and well-estimated architectonic element. Examples are Brandts in Odense, Nicolai in Kolding, Papirfabrikken in Silkeborg and Nordkraft in Aalborg. The "container" is a term developed to describe cultural projects that in a relatively anonymous manner slides into the context but which in no way calls for attention architecturally speaking. Neither do such projects work as an architectural force. It can be newer buildings or it can be re-developed older buildings. Examples are Underværket in Randers and Grey Hall & Loppen (Christiania). The final category is "temporary architecture", and as the name suggests, these projects are predominantly characterized by their ephemeral and non-permanent features. They might be installations or temporary events, which are either unique and only exist once or they may be re-created in a cycle of temporary creation. The temporary architecture projects may have large branding implications or they might direct attention towards overlooked architectural qualities and potentials in the city. Examples of this category are Amagerbogade and Metropolis in Copenhagen, the Festival Week in Aarhus or Middle Age Days in Horsens.

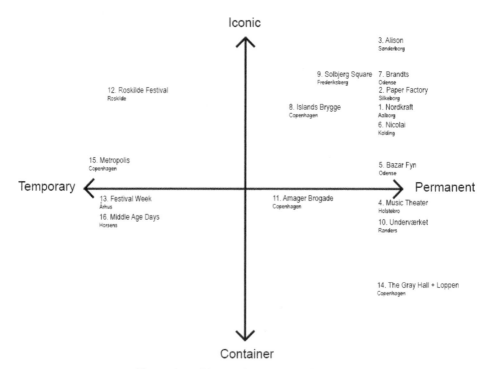

Figure 6. Architectural status (at project start)

Finally, we plotted the projects onto a map illustrating the architectural status in the urban fabric with the axis "temporary–permanent" and "iconic–container" spanning the analytical space (Figure 6). The result is a map of projects clustered in the field "iconic–permanent". However, there are identifiable clusters of "container–permanent". What might be said to be surprising here is the relatively low proportion of temporary and container types, suggesting not only permanence but also substantial political priority and economic commitment.

5. Discussion and Concluding Remarks

In the discussion and concluding remarks, we would like to return to a more general discussion. That is to say, when trying to understand the design of the experience city, we need to point to issues of the institutional make-up and the configuration between, on the one hand, the new urban agenda and, on the other hand, the new planning strategies, the change in the social and material practice and the use of architectural narratives and form. We need to enquire into what are the prevailing rationales underpinning the projects and interventions. There are important issues of how the interventions facilitate material practices and symbolic interpretations and how the relation is between the use value and issues of symbolic and experiential value. As our understanding of the experience city is related to the social geographies of power these interventions shape, we are interested in which types of interaction and experiences are enabled (or constrained) by these urban interventions. Furthermore, we also enquire into which types of interactions and mutual learning processes will be shaped in and by the projects. Our findings exploring the Danish experience city point at new ways of conceptualizing the city.

In relation to urban transformation and planning/strategic planning, we found four main categories: lighthouse, linear attractor, field and void. In relation to private–public actors, there were a large cluster of projects at the "lighthouse–public" field and only one in the "private–void" area, suggesting a strong political lead in the public projects. The "elite–many" pole suggests quite some variation in the types of projects and a small cluster at the "few–elite" pole. In relation to the project's role as creator of hybrid cultural programmes, we found that most projects had many and different programmes. This related to the main target of cultural elite and cultural public. The categories of target groups were cultural elite, cultural avant-garde, cultural public and subcultures. Mapping these onto a diagram with an axis spanning the model "many–few" programmes showed that only a few projects considered the subcultures. Finally, when looking at the relationship to physical transformation and architectural expression, we found the four categories of icon, historic monument, container and temporary architecture. Plotting the projects onto a diagram illustrating the architectural status in the urban fabric with the axes "temporary–permanent" and "iconic–container" spanning the analytical space results in a map of projects clustered in the field "iconic–permanent", suggesting a certain political and economic prioritization of the projects.

The results presented in this article do not show the full variation of the complex relations, so the presented general overview of projects will be followed up with detailed case studies. However, on a general note, some preliminary conclusions are possible. The intent behind the new projects is to give the urban culture a qualitative lift through a combination of different programmes that are expected to further creativity and artistic exertion. They are also expected to help promote fellowship and understanding between many different cultural groups in the city. The cultural life in the city can be strengthened in a socially sustainable manner. Urban culture is put on the agenda as many different issues at the same time: a culture of knowledge and learning, a physical culture for play and performance, a tolerant culture for the social encounter and a participatory entertainment culture that reaches far beyond street musicians and café latte. The projects intervene in city life and the citizens' use of the city's venues and available cultural offers. Initially, many of the projects seem to be established at disused industrial facilities and to borrow their scale and typological multiplicity from the industrial architecture and the harbour environments while simultaneously adding new architectural elements of their own. The projects are very attentive to architectural and spatial transparency, transition and open programming.

From the analysis so far, we have partly found reason to establish new concepts in order to come to grasp with these new hybrid forms of urban interventions in the experience city. Furthermore, we have shown some of the differences and similarities among the cases. However, more knowledge is needed in order to gain a deeper understanding of these new urban interventions and practices. In particular, we would argue that this is the case within three more distinct fields, that of urban transformation and planning strategy, urban culture and architectural typologies and urban spaces. Research into the experience city will form a considerable basis for politicians, urban designers and private investors in their planning of the urban environments of tomorrow. It will, however, also ensure that the citizens' experiences with new urban qualities and designs are considered in order to create unforeseen and culturally enriching city experiences rather than merely commercial experiences. The production of new knowledge within this field of research will be vital for overcoming the new urban challenges that confront the experience city in the longer term.

References

Bauman, Z. (2002) City of Fears, City of Hopes, in: H. Thomsen (Ed.) *Future Cities: The Copenhagen Lectures*, pp. 59–90 (Copenhagen: Fonden Realdania).

Boer, F. & Dijkstra, C. (2003) Funscapes: The European leisure landscape, in: R. Broesi, P. Jannink, W. Veldhuis & I. Nio (Eds) *Euroscapes*, pp. 167–214 (Amsterdam: MUST Publications).

Braae, E. (2003) *Konvertering af ruinøse industrilandskaber* (Converting Brownfields), (Aarhus: Aarhus School of Architecture).

Bunchoten, R. (Ed.) *Urban Flotsam: Stirring the City* (Rotterdam, NY: 010 Publishers).

Castells, M. (1996) *The Information Age: Economy, Society and Culture, Vol. 1: The Rise of the Network Society* (Oxford: Blackwell).

Cullen, G. (1971) *The Concise Townscape* (Oxford: Architectural Press).

Evans, G. (2001) *Cultural Planning: An Urban Renaissance?* (London: Routledge).

Florida, R. (2002) *The Rise of the Creative Class—and How It's Transforming Work, Leisure, Community and Everyday Life* (New York: Basic Books).

Gehl, J. & Gemzøe, L. (2000) *New Urban Spaces* (Copenhagen: The Danish Architectural Press).

Goffman, E. (1963) *Behaviour in Public Places: Notes on the Social Organisation of Gatherings* (New York: The Free Press).

Greenberg, M. (2000) Branding cities: A social history of the urban lifestyle magazine, *Urban Affairs Review*, 36(2), pp. 228–263.

Hajer, M. & Reijndorp, A. (2001) *In Search of New Public Domains* (Rotterdam, NY: NAi Publisher).

Hall, P. (2000) Creative cities and economic development, *Urban Studies*, 37(4), pp. 639–649.

Hall, T. & Hubbard, P. (Eds) (1998) *The Entrepreneurial City: Geographies of Politics, Regime and Representation* (Chichester: John Wiley).

Harvey, D. (1990) *The Condition of Postmodernity: An Inquiry into the Origins of Cultural Change* (Oxford: Blackwell).

Jensen, O. B. (2007a) Culture stories: Understanding cultural urban branding, *Planning Theory*, 6(3), pp. 211–236.

Jensen, O. B. (2007b) Brand resistance and counter branding, in: G. Marling & M. Zerlang (Eds) *Fun City*, pp. 99–120 (Copenhagen: The Danish Architectural Press).

Jessop, B. (1997) The entrepreneurial city: Re-imagining localities, redesigning economic governance, or restructuring capital? in: N. Jewson & S. MacGregor (Eds) *Transforming Cities: Contested Governance and New Spatial Divisions*, pp. 28–41 (London: Routledge).

Jessop, B. (2004) Recent societal change: Principles of periodization and their application on the current period, in: T. Nielsen, N. Albertsen & P. Hemmersham (Eds) *Urban Mutations. Periodization, Scale, Mobility*, pp. 40–65 (Aarhus: Arkitektskolens Forlag).

Kendahl, S. C. (2003) Wien has become the suburb of Graz, *Information*, (18.01.03, Copenhagen).

Kiib, H. (2006) Kreative Videnbyer, *Byplan*, 58(3), pp. 112–117.

Kiib, H. (2007) *Harbourscape* (Aalborg: Aalborg University Press).

Klingmann, A. (2007) *Brandscapes: Architecture in the Experience Economy* (Cambridge, MA: MIT Press).

Kunzmann, K. (2004) Culture, creativity and spatial planning, *Town Planning Review*, 75(4), pp. 383–404.

Kvorning, J. (2004) Urban restructuring and cultural planning, in: K. Østergaard (Ed.) *Cultural Planning*, pp. 53–60 (Copenhagen: Center for Urbanism, The Royal Academy of Fine Arts).

Landry, C. (2000) *The Creative City: A Toolkit for Urban Innovators* (London: Earthscan).

Lash, S. & Urry, J. (1994) *Economies of Signs and Space* (London: Sage).

Lever, W. F. (1999) Competitive cities in Europe, *Urban Studies*, 36(5/6), pp. 1029–1044.

Lynch, K. (1960) *The Image of the City* (Cambridge, MA: MIT Press).

Marling, G. (2003) *Urban Songlines—Hverdagslivets drømmespor* (Aalborg: Aalborg University Press).

Marling, G. & Kiib, M. (2007) Designing public domain in the multicultural experience city, in: H. Kiib (Ed.) *Harbourscape*, pp. 106–113 (Aalborg: Aalborg University Press).

Marling, G. & Zerlang, M. (2007) *Fun City* (Copenhagen: Arkitektens Forlag).

Metz, T. (2002) *FUN. Leisure and Landscape* (Rotterdam, NY: NAi Publishers).

O'Dell, T. & Billing, P. (Eds) (2005) *Experiencescapes: Tourism, Culture and Economy* (Copenhagen: Copenhagen Business School Press).

Pine, B. J. & Gilmore, J. H. (1999) *The Experience Economy: Work Is Theatre and Every Business Is a Stage* (Boston, MA: Harvard Business School Press).

Rogerson, R. J. (1999) Quality of life and city competitiveness, *Urban Studies*, 36(5/6), pp. 969–985.

Schulze, G. (1992) *Die Erlebnisgesellschaft: Kultursoziologie der Gegenwart* (Frankfurt: Campus).

Thackara, J. (2005) The post-spectacular city—and how to design it, in: S. Frank & E. Verhagen (Eds) *Creativity and the City. How the Creative Economy Changes the City*, pp. 184–191 (Rotterdam, NY: NAi Publishers).

Zerlang, M. (2004) The cultural turn in contemporary urban planning, in: K. Østergaard (Ed.) *Cultural Planning*, pp. 7–10 (Copenhagen: Centre for Urbanism, The Royal Academy of Fine Arts).

Experience Spaces, (Aero)mobilities and Environmental Impacts

CLAUS LASSEN, CARLA K. SMINK & SØREN SMIDT-JENSEN

ABSTRACT *This article investigates how aeromobility is used as a core element in the development of new urban strategies of experience and transformation of urban spaces. Two examples have been selected and studied: the municipalities of Billund (Denmark) and Nyköping (Sweden). It is argued that both examples are not just showing a simple form of causality, where increased access to air travelling creates a new experience destination. They also illustrate the complex impact of the increasing prevalence of air travel on the spatial, social and economic development of the cities, and at the same time, how the spatial, social and economic reorganization contributes to the prevalence of air traffic, airports and air spaces.*

Introduction

In the past few decades, the number of experience opportunities has exploded, and experiences seem to have become increasingly important in modern life. The experience economy focuses on the rising demand for experiences in society, and it builds upon the added value of creativity generated both in new and traditional products and services (Bærenholdt & Sundbo, 2007; Lorentzen, 2009; Pine & Gilmore, 1999). It is a type of economy that capitalizes on a consumer society in which money is spent increasingly on leisure, arts and cultural events. However, the notion of experience economy can also be seen as a way to understand and describe a new stage in modern capitalism towards a more imaginative and symbolically oriented economy of production (Shaw & Williams, 2004, p. 116). In several European cities, experience now has an increasing influence on urban politics, urban planning and urban designs (Hall, 2000; Kunzmann, 2004; Metz, 2002; Marling *et al.*, 2009).

This article focuses particularly on aeromobility[1] in relation to the discussion of experience economy. As Whitelegg (1997) points out, the drive to consume large distance, as part of the search for experience, reaches its apogee in global tourism and air travel. Today, flying is a fundamental element in the process of economic and cultural globalization (Graham, 1995); and globally there are 1.9 billion air journeys each year (Urry, 2007, p. 150). Increasing air travel is not only transforming work and family life in the western world, but also the possibilities of experience and leisure. This is particularly related to the increasing number of cheap air tickets and the number of low-price destinations in Europe. In 2006, approximately 140 million passengers were carried by low-fare airlines within Europe to more than 280 airports (European Low Fares Airline Association, 2007, p. ii). Air travel not only extends the possibilities and scale of leisure consumption on an individual level, it also functions as a tool to construct, transform and brand "places of experience", as will be illustrated in this article.

However, using air traffic as an important tool in the production of experience is problematic. Due to a more aggressive impact of CO_2 emissions in the higher strata of the atmosphere, the threat to the global climate from airplane emissions has become more serious than the threat from emissions of vehicles, which travel at the same distances at surface level (Høyer & Næss, 2001; Lassen, 2005; Gössling & Peeters, 2007). Today, the rapidly expanding air traffic worldwide contributes about 3% of the production of CO_2 to the global climate (Frändberg, 1998, p. 73). One other important consequence of this increase in air transport is that tourism now accounts for more than 60% of air travel and is therefore responsible for an important share of air emissions (www.uneptie.org). The local impact of air transport on the ground is also significant and includes: land take for airports, terminals and runways; noise and air pollution from aircrafts; pollution from buildings; air pollution and noise from roads and road transport serving the air transport (Whitelegg, 1997, p. 86).

Through the two examples of Billund (Denmark) and Nyköping (Sweden), this article explores how aeromobility is used as a core element in the development of new urban strategies of experience and transformation of urban spaces. The article focuses on the complex relations between the urban/regional growth strategies, experience economy and (aero)mobility. The examples have been studied as case studies involving qualitative research interviews, literature reviews and statistical material as data-collecting techniques. The two examples have been selected because they are considered to be cases of the complex connections between increasing aeromobility and social, economic and spatial reorganization. It is argued that Billund and Nyköping are not just picturing a simple form of causality, where increasing access to air travel creates a new experience destination; the two examples also illustrate the complex impact of the increasing prevalence of air travel on spatial, social and economic development of the cities, and at the same time, how spatial, social and economic reorganization contributes to the prevalence of air traffic, airports and air spaces. As such, the article also discusses how new forms of hypermobility (Adams, 2005), which are connected with transformed spaces for leisure and play, are a big challenge to politicians and planners on various levels from the local to the global in terms of environmental and climate change problems.

The article is divided into four sections. In the second section, we explore the two examples, where the relationship between growth of the local/regional experience economy and aeromobility has been a major concern. In the third section, we discuss the examples in the light of the notion of sustainable development, low-price aeromobility and experience economy. Finally, the conclusion puts into perspective some of the planning and policy implications to be extracted from studying the two examples.

The Billund Case: From Industrial Area to "Destination of Experience"

The first example, which we argue can be understood through the notion of experience economy and aeromobility, is Billund in Denmark. The municipality has 26,000 inhabitants (2007) and is located in central Jutland in Denmark (Figure 1).

Historically, Billund has been characterized by a large number of industrial jobs and high traditional industrial production, especially bacon factories and the production of toys (home of LEGO). The transformation of the economy and labour market has recently caused a high number of industrial jobs to be outsourced to less-developed countries with lower wages or closed down. During the last decades, the area (within the municipal borders that include both the city and its near surroundings) has lost a large number of "blue collar" jobs, and 10% of the total number of such jobs have disappeared during the last 2–3 years alone, and today the area is facing a period of economic decline (Director of Billund Business Association, 2007). Due to this development, the municipal council decided to focus on the experience economy, and consequently they made a strategy for economic development with a starting point in experiences and tourism. The purpose was to create a local labour market where more people would be employed within the tourism sector.

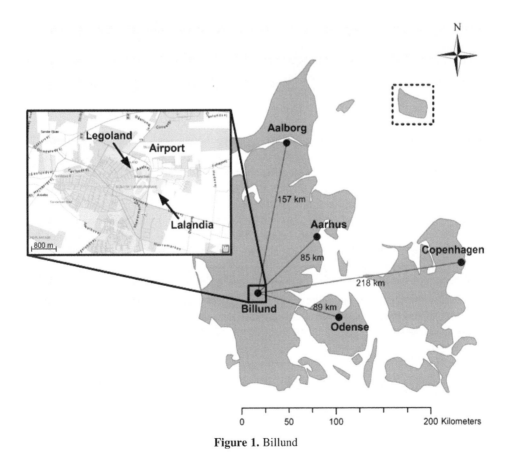

Figure 1. Billund

The new strategy in Billund has been established and driven forward by a few large national and global players as well as the municipality and the local business association. This was a rather limited combination of actors, compared with the often broader approaches applied in other medium-sized Danish cities (see Lorentzen, 2009; Therkildsen *et al.*, 2009). The new urban growth strategy has resulted (physically) in a number of building activities reshaping the urban environment in the area surrounding the airport, and a number of investors have invested in local projects.

The experience strategy in Billund has primarily been concerned with the development of Legoland, Lalandia Billund (under construction) and Billund Airport (Figure 1). "Legoland" is an amusement park based on the products from the international toy company LEGO. Similar parks can be found in London, Berlin and California, and today they are all owned by Merlin Entertainments Group, which is a global actor in the industries of entertainment, pleasure and experience. Merlin Entertainments Group has 51 attractions, which employ over 13,000 staff and attract more than 30 million visitors in 12 countries across Europe, North America and Asia (www.lalandia.dk/billund). The goal of "Lalandia Billund" is to establish one of the greatest experience and holiday centres in Northern Europe for families with 40,000 m^2 for visitors from both Denmark and abroad (www.lalandia.dk/billund). Lalandia is owned by PARKEN Sport & Entertainment, which is one of the most important national actors inside the sector of sport and entertainment (www.parken.dk). For example, the company runs FC Copenhagen, financially the largest football club in Scandinavia, organizes concerts and events as well as centres of leisure, play and holidays. Finally, the international airport of Billund is an important actor. In 1962, LEGO took the initiative to establish this airport. The initiative was supported by the county of Vejle and the local city councils and parish councils. In 1964, Scandinavian Airlines (SAS) established regular flights between Billund and Copenhagen, and in 1966, multiple charter flights to tourist resorts started out. Billund Airport is now the second-largest international airport in Denmark with 1.98 million passengers per year (2005). The airport has domestic flights to Copenhagen and Rønne (Bornholm), 29 European routes and 1600–1700 people are employed. The airport is owned by several municipalities in area, but it is operated as an independent business (Director of Billund Airport, 2007).

The three aforementioned actors have played key roles in developing the experience strategy in Billund together with the municipality and the local business association (Director of Billund Business Association, 2007). Historically, tourism in Billund has primarily been based on transit passengers, who only stopped for a short period of time on their way to other places and for day trips to Legoland (1.6–2 million visitors each year), which is located next to the airport (Figure 1). A key element in the development of the new strategy has been the wish of the local business association to get more value out of the transit passengers and day trip tourists as well as to attract new visitors to the area. However, talks about creating a new urban strategy first became serious in 2006 when a number of external actors wanted to invest in a new centre of entertainment for families, and when Merlin Entertainments Group at the same time took over the amusement park Legoland. This meant that the discussions and ideas were now transformed into real action:

> Together this meant that some activities, which gave trust in the economic deve-
> lopment, had started. In Legoland, a number of new activities had started, and

the airport believed that new and more destinations could be established. All the different activities meant that the vision of making the region an experience municipality, or an experience region, was suddenly visible to everybody. There were suddenly some money, machines, and some land behind the ideas. It helped making the whole thing move. (Director of Billund Business Association, 2007)

Another important element, which contributed to push the process forward, was that Billund municipality (the municipal council and administration) made the necessary plans and reservation of areas that made it possible to accomplish the project. It can be concluded here that the municipality, through a variety of both local/regional, national and global players and forces, has been trying to facilitate a spatial transformation towards reshaping the area into a destination of fun, play and experience. As we will show in the following, this transformation has been closely linked to the local airport and low-price air flights.

From Mass Tourism to Individual Planning of Experience

A key element in the strategy of experience in Billund is aeromobility, and the local airport is therefore a central player. As part of the local/regional growth strategy, the airport has stated that it wants to attract more low-price flights to and from Billund Airport. The goal is to attract more tourists and visitors to the area in the future, and the expectation is that in 2011, more than 5 million people will visit the place, of which 80% should be foreigners (Director of Billund Business Association, 2007). In the light of this goal, the attraction of low-cost air companies plays an important role. Alongside the attraction of more visitors to the Billund area, an important part of the strategy is to function as a port to tourist sites and attractions in Western Denmark (Director of Billund Airport, 2007).

Historically, the airport has been based on charter traffic out of Billund Airport, but recently the number of international route traffic flights has increased significantly. Since the opening of a new passenger terminal in 2002, the airport has had capacity for 3.5 million passengers a year. In the period 1990–2004, the number of passengers increased by 650% (www.billund-airport.dk/). The last 3 years, a number of European cities have been connected to Billund by direct flights, and the increased supply of international routes provides a very important material contribution to the creation of an experience economy in Billund. The actors involved in the project propose the hypothesis that there exists a new type of "experience travellers" who can be attracted due to increased international access to the area of Billund:

> One of the things that we have to be very aware of is that we have a very large group of tourists who plan with one day's notice. They say: We have a weekend off; how should we spend it? We could go to Billund, it costs 2 Euro plus airport tax with Ryanair to Billund Airport. . . .Here we can see that more Spaniards are flown in to Billund airport than to Copenhagen Airport. Ryanair comes from Spain several times a week and on these flights there are in fact more Spaniards than Danes. (Director of Billund Business Association, 2007)

The Billund Business Association, the local business collaboration, presumes that a new type of traveller is becoming interested in Billund. Apart from cheap flights, another element in attracting such a new type of tourist is the use of virtual mobilities on the Internet, where the municipality and other local actors are trying to brand Billund as a city of experience. The goal is to offer a homepage for Billund in the languages of the countries from where it is possible to fly to Billund airport. Low-price tickets and Internet marketing in various languages are orientated towards the new kind of travellers. In particular, the actors involved have defined the new type of experience travellers as people coming from European cities (Malaga, Madrid, Barcelona, etc.), who, contrary to traditional mass tourists, do not make plans before leaving their homes, but instead they plan while on the trip (see Urry, 1990). This type of traveller discovers Billund on the Internet as a possible place for a "weekend of experience and play", and only when they arrive will they book a hotel in the area of the airport from where they will plan the rest of the journey. Such travellers make up the programme themselves: they buy the air tickets themselves, they book the hotel themselves and they buy the ticket to tourist sites and events themselves. Tourist and leisure travel is no longer something that is planned for months in advance; instead, it is put together based on the available options. Whether there exist in reality such new segments of tourist travellers or not, it is remarkable that the formulation of the experience strategy and a number of the urban initiatives taken in Billund are based on the assumption of such a new type of air traveller.

Low-Price Air Companies and International Travel as the Main Tools

In the attempt to create a more eventful place for people to visit, the attraction of low-price companies and new low-price routes has, so far, played an important role. Billund airport wants to make the best possible use of the capacity related to the increasing route flights, so that Billund is not only a place of departure, but also a destination. The low-cost airline company Ryanair has upgraded Billund airport with flights to four new international destinations in 2007, and from 2008 onwards Ryanair wants to have flights to seven international destinations (ErhvervsBladet, 2007). Likewise, the Danish low-cost company Sterling has recently established a number of new routes, and the full service air company KLM has selected Billund as a new Scandinavian hub in their network strategy with an increased number of flights to Schiphol Airport (Amsterdam). The recent development of a more experience-orientated economy in Billund is closely related to the increasing number of low-price flights from abroad to Billund airport. From Figure 2, it appears that the inbound leisure traffic has increased in 2006 and 2007, and it is also expected to increase further in 2008. To a large extent, this development is generated by a significant increase in the number of low-price passengers.

The Nyköping Case: Building Aeromobility and Experiences into a New City Brand

Another example which can be understood through the notion of experience economy and aeromobility is the city of Nyköping, with 50,000 inhabitants and located on the Baltic coast of Sweden coast approximately 100 km south of central Stockholm (Figure 3). Nyköping is an interesting example because, during the last 8–10 years, the city has developed a set of new urban strategies that has included (aero)mobilities and experiences as two key elements. As we shall explain below, these strategies have to a large extent been interweaving.

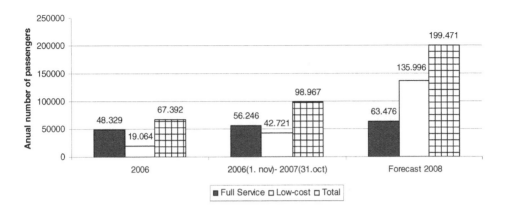

Figure 2. Inbound leisure traffic through Billund Airport 2006–2008
Source: Billund Airport (2007)

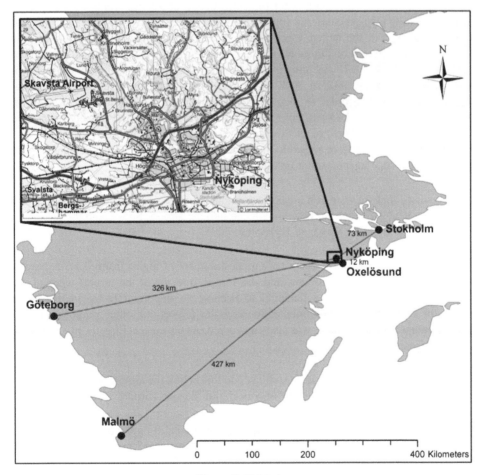

Figure 3. Nyköping

From an Anonymous City to a Hub of Mobilities

In a historical perspective, Nyköping has had many jobs in the sector of public administration, and until the mid-1980s the city was also an important industrial location in Sweden due to the production of ships, automobiles, textiles, quality furniture, televisions and soap (Groth *et al.*, 2005). In the late 1990s, most major industries had left the city, although with no major wounds neither economically nor to the physical environment. The main role in sight was to be "just an average Swedish city" (Groth *et al.*, 2005, p. 164). Speculations emerged that it could become an attractive residential city for people working in the greater Stockholm region, although the distance was relatively far, either by taking the E4 or the bumpy regional train (Mayor of Nyköping, 2005). However, as Ryanair in 1997 opened their first international route from London-Stansted to Stockholm Skavsta, it became clear to the municipality of Nyköping (the municipal council) that the city perhaps also could have a role as a hub of aeromobilities.

The Stockholm Skavsta airport was taken over by the municipality of Nyköping and neighbouring municipality Oxelösund from the Swedish Armed Forces in 1984, and was changed into a civil airport. After the construction of a longer runway in 1990, the airport started to take on chartered flights. However, until the mid-1990s the airport had limited success, and the deficit was covered by the two municipalities. As it became evident that the success of flights by Ryanair between London-Stansted and Stockholm Skavsta (initiated in 1997) was immense, and in the light of the deregulation of the European airline market in 1997, in 1998 the municipalities decided to sell the majority of shares in the airport in order to secure the best conditions for expansion. As a result, 90% of the airport stocks were sold to an UK-based investor TBI closely connected with Ryanair (Current Director of Stockholm Skavsta Airport, 2003). Prior to the sale, there had been an extensive redevelopment of the facilities at the airport, paid by the municipality, a state agency and a private developer:

> We realised that we were at a breaking point and that we needed to have new management, new investors and new knowledge into the airport. So we sold 90 percent of our stocks to TBI. (Mayor of Nyköping, 2005)

Prior to 1997, there had only been a very modest number of flights from Skavsta. Nevertheless, the municipality politicians and the former director of the airport were of the opinion that the airport had a potential to become an important factor in the future growth of Nyköping. At the time, converting the airport for other purposes, e.g. a housing area or a business park, was also considered. As the current director of Stockholm Skavsta Airport formulates it:

> Had it not been for this long term strategy and this belief, the airport would not have been here today. (Current Director of Stockholm Skavsta Airport, 2003)

The relationship between Ryanair and Nyköping (municipality) grew even stronger when a 10-year co-branding agreement was signed in 2003, worth 55 million SEK (approximately 5.5 million euro) between the two partners. By paying this amount to Ryanair over 10 years, the municipality was promised exposure of the municipality's name and logo on the outside of airplanes flying to Stockholm Skavsta, exposure on the Ryanair website with a banner that linked to the official website of Nyköping municipality and exposure of the municipality's name and logo together with Ryanair in newspaper adds, radio and TV commercials, etc. However, the deal with Ryanair was strongly debated and much criticized when it was signed in 2003 without a public tender process. Nyköping municipal council, led by a highly entrepreneurial social democratic mayor (Göran Forssberg), was accused of hidden public support to a private company and that the agreement was a trick to make Ryanair choose Stockholm Skavsta airport as their entrance to Stockholm.

Despite the controversy about the alleged support by the municipality, and after three different courts had decided that the agreement was against the law, the success of the Stockholm Skavsta airport and Ryanair was massive: the number of travellers increased by 1448% in the period 1997–2007 (from 128,862 in 1997 to 1,995,000 in 2007), which made Skavsta the third-largest airport in Sweden. The major share of travellers was Ryanair customers (1.6 million of 1.8 million travellers in 2006). Tourism in Nyköping was booming as well. The number of full-time employment in tourism increased from 448 in 2002 to 757 in 2007; the number of hotel overnights increased from 107,000 in 2002 to 139,000 in 2005. The total turnover in tourism in Nyköping was estimated to be 572 million SEK (approximately 57 million euro) in 2002, while it had increased to just above 1 billion SEK (approximately 100 million euro) in 2007 (http://nd-vip1.news-desk.se/pressroom/nykoping/pressrelease/view/96368). Similar to Billund, Nyköping has also attracted a new type of experience travellers. What distinguishes Billund from Nyköping is that the airline travellers coming to Nyköping are almost entirely travellers that come with low-cost carriers, since Stockholm Skavsta predominately services such airline companies.

In October 2007, Ryanair opened 13 new destinations from Nyköping, primarily to Southern Europe (Alicante, Basel, Berlin, Bratislava, Eindhoven, Baden-Baden, Liverpool, Malta, Pisa, Oporto, Salzburg, Trapani and Valencia) (www.skavsta.se). With the new destinations, Ryanair has 28 destinations from Stockholm Skavsta out of 42 destinations for the entire airport. The other airlines servicing Stockholm Skavsta are Wizz Air, Gotlandsflyg, Fritidsresor, Ving and Apollo. The new Ryanair destinations are expected to create 100 additional jobs in Nyköping, bringing the number of employed persons at Stockholm Skavsta up to approximately 800 persons. In 2003, before the explosive growth took off, about 100 people were employed at the airport (Current Director of Stockholm Skavsta Airport, 2003). With the new destinations, 180 employees at Ryanair will have Stockholm Skavsta as their base. In addition, new investments will be made at the airport to make the terminals, shops, parking, etc. suitable for the increasing number of expected travellers.

The future plan is to increase the number of travellers to 6 million within a few years, and in 2007 environmental permissions were given by the Swedish national authorities for this number of travellers. A precondition, which has been pointed out by the airport as well as local decision-makers, is that the Swedish government decides to build a fast rail between Stockholm and Norrköping, going through Nyköping and Stockholm Skavsta airport, the so-called Ostlänken. Since the mid-2000s, Nyköping has lobbied massively for this project because it would increase the attractiveness of the city as a place to commute from, but also in order to improve the access to Stockholm Skavsta airport.

Increasing Demand for Experiences

The advertising agreement between Ryanair and Nyköping, signed in 2003 (updated 2007), made it possible for the partners to set up mutual demands. In 2005, Ryanair explicitly demanded that Nyköping should provide experiences for travellers going with Ryanair to Stockholm Skavsta. Ryanair demanded that these activities (e.g. various activities for visitors and cultural events) should change every week to make it possible for travellers to experience something new at each visit. As the municipality was prohibited, according to Swedish law, to initiate commercial activities themselves or to support single economic interests, the municipality had to request and engage the local business life into being more active in order to live up to the agreement with Ryanair. This demand led directly to the birth of a new organization "Visit Nyköping", which is funded by local private companies and organizations involved in tourism, and an adjoining Internet portal with a comprehensive overview of the local sights, activities and events. As long as the initiative came from Visit Nyköping, the municipality was prepared to participate and finance at least a part of the activities as well. The local business life could make initiatives that the municipality would then support by providing the best conditions for the realization of the initiatives. The demands from Ryanair also led to increasing cooperation with surrounding municipalities in the region (Halldin & Hultman, 2005).

The success of Ryanair and the Stockholm Skavsta airport has also been important in transforming the city both discursively and materially from "a city at the periphery of Stockholm" to "an integrated member of greater Stockholm" (Mayor of Nyköping, 2005). This new turn has been reflected in the fact that an increasing number of people, especially young and wealthy seniors, are moving to Nyköping, particularly from Stockholm (Smidt-Jensen, 2005, p. 3). Those groups have in particular been targeted in a massive promotion campaign conducted by Nyköping since 2003 in order to attract new settlers. Residents in Stockholm have especially been targeted, as housing prices in Stockholm have grown dramatically. The anonymous city image of Nyköping—as a civil servant city with standard Swedish welfare and urban amenities—has been transformed into a fresh brand based on three core qualities of the city: "the good life", "the sea" and "limitlessness". This spirit of Nyköping has been communicated via a few key images. For example, the city logo pictures two young happy people on a scooter, one person on the sea canoeing and a departure sign from the airport saying: Paris, Brussels, Hamburg, Katowice, Milan, Frankfurt, Budapest, London and Glasgow. This illustrates that Nyköping not only tries to brand itself as a place of safety, happiness and joy, but also as an international hub where it is easy for modern people to connect to the surrounding world of excitement and experience. The underlying message of the city logo is: "If you move to Nyköping, you can be close to nature, live and work in greater Stockholm, and travel easily the world". This image has been used in Swedish media as, for example, in Stockholm newspapers, lifestyle magazines and websites for housing and living. Here aeromobility is used as an important ingredient in the construction of the new Nyköping, and in doing so the development of Stockholm Skavsta airport and the low-cost air company Ryanair are fundamental material elements in fulfilling this vision. Most of the promotion activities are made by the municipality's own marketing department set up in 2002, one of the first in a Swedish municipality, and with a budget of 5–6 million SEK per year. The establishment of this department underlines the high priority that the city promotion has received, and it shows the new conditions of the inter-urban competition for Swedish municipalities in the larger Stockholm region:

> Formerly it was very seldom that you saw a campaign or an advertisement for a Swedish city promoted by a municipality. It was about "running business" and distributing welfare to the inhabitants. Of course there was information material available for tourists. Today we have a marketing-department at the city hall with a rather big budget. They work with establishing Nyköping as a brand in the Greater Stockholm area. (Mayor of Nyköping, 2005)

In order to have attractive housing offers for new residents, the municipality has been pushing and planning for the use of new land for residential purposes in highly attractive areas on the coast of the archipelago, and the former industrial part of the harbour area has been converted by a private contractor into attractive industrial-style housing. A large publicly funded urban renewal project in the harbour area has also been undertaken, where former storage houses have been turned into chic restaurants, bars and cafes. In addition, a large dock primarily for guest boats has been constructed in connection to the old dock, a place that is also used by, for example, German and British visitors flying to Nyköping, where they have their private boat and access to the archipelago (Current Director of Stockholm Skavsta Airport, 2003). The municipality has also been facilitating the construction of the "Rosvalla Nyköping Event Center", a large multi-arena for national sports and music events. The cultural event in Rosvalla that has had the most massive exposure in Swedish media so far was the Swedish Eurovision song contest qualifier in 2007.

Those projects have been promoted in order to make Nyköping attractive for new settlers with an everyday life connection to the Stockholm region. However, they can also be related to the process of changing the city into a destination for the many new travellers coming by air to Nyköping. For many years, the city passively counted on the archipelago and the Nyköping castle as its primary qualities, but since Ryanair chose to make Stockholm Skavsta its Scandinavian hub, and after the number of visitors increased substantially, the situation has changed (Mayor of Nyköping, 2005). As mentioned earlier, the co-branding agreement with Ryanair has made it possible for Nyköping to get widely promoted in other Ryanair destinations. And lately, the attempts to target potential visitors from German Ryanair destinations has included offering "Inga Lindström-trips" named after the leading female character in a highly popular soap opera broadcasted in Germany, which takes places in Nyköping and its surroundings (Nyköping Kommun, 2007). The strong increase in overnights in Nyköping hotels is another indication that Nyköping has been turning into a true destination.

The Transformation of Urban Spaces and the Increasing Low-Price Aeromobility

In this section, we will discuss some of the implications related to experience growth strategies based on the increase of (low-price) aeromobility. As it appears above, Billund and Nyköping can be seen as cases of the development of experience economy linked together with the attraction of (low-price) airlines. The airports in the two cities and the increased access to low-price air tickets have also been important in relation to the development of both Billund and Nyköping into more eventful destinations and attractive places to live. The cities have attempted to change their images from anonymity into a location of experience connected to the global tourism flows. Local investments, policies and planning have to a large extent been directed towards the creation of an experience dimension, which contributes to providing airlines and international travellers a concrete purpose for flying to the cities. Without the experience dimension it would not be an obvious choice to travel to a place which they have never heard about before, and which does not appear to have any special or unique tourist experiences, but rather appears as trivial and anonymous for visitors. In this context, it is important to understand that there is a close connection between the prevalence and growth of a particular form of mobility and the social reorganization which takes place in relation to this particular type of transportation. Especially, the travel industry is the organizer of the modern experience (Urry, 1995, p. 142). The aviation industry can be seen as a tightly connected system of places, private corporations and state actors, interrelated with almost all other sectors of the economy (Urry, 2007, p. 148). In this perspective, it can be argued that airplanes are nothing without airspaces, and airspaces are nothing without "the impeccable machine making use of its splendid expanses" (Pascoe, 2001, p. 21).

In the cases of Billund and Nyköping, it means that reorganization of the local social organization in form of developing experience strategies plays an important role in relation to increasing incoming low-price air travel. The examples from Billund and Nyköping are not just showing a simple form of causality, where increased access to air travelling creates a new experience destination; the examples illustrate the complex impact of the increasing prevalence of low-price air travel on the spatial, social and economic development of the cities, and at the same time, how the spatial, social and economic reorganization contributes to the prevalence of air traffic, airports and air spaces.

Both cases show how a more imaginative and symbolical economy not only results in new local movement patterns, but they also show how different scales and mobility practices as well as horizons of meaning are interwoven globally. For example, Nyköping not only tries to launch concrete experience initiatives, but the municipality also tries to connect the city with the global space of flows (Castells, 1996) at both a material and discursive level by working with the development of the airport and marketing the city as a place with local experience and international access to European metropolises through low-price air routes. In Billund, the focus is also to create urban initiatives that can attract international visitors through investments in old as well as new centres of experiences, combined with international marketing on the Internet, and at the same time, the local airport is developed both as a low-price airport and as a network airport. Both examples therefore illustrate how different scales and mobility practices are involved and combined in the local urban spatial strategies and policies.

Global Perspectives on Aeromobility and Environment

However, the transformation of urban spaces, with a focus on attracting more global tourists and visitors, seems problematic on different levels. One general characteristic of aeromobility is that over the past 45 years it has changed from a luxury form of mobility for the wealthy few into a contemporary form of hypermobility. This hypermobility is characterized by promises of cheap high-speed travel and the inclusion of new social groups in air transportation, including the mass movement of long-distance tourists

(Gössling & Peeters, 2007, p. 403). In many ways, it can be argued that air travel has become the industry that stands for, and represents, the new global order (Urry, 2007, p. 135). Whitelegg (1997, p. 77) identifies several mechanisms behind this development:

> One of the most important areas in which global air travel has affected massive social and economic change has been that of increasing the accessibility of remote and fragile areas to the global tourist. The increase in accessibility, coupled with higher levels of disposable income, more leisure time and better education have promoted the demand for foreign holidays; furthermore, this increase in demand has enabled tour operators to keep prices low, thereby satisfying this demand with offers of relatively cheap package tours and charter flights to more remote and exotic locations.

Obviously, those developments are in conflict with goals to achieve environmental sustainability. In general, environmental impacts from air transport can be summarized to be effects from greenhouse gases (CO_2 amongst others), hydrocarbons, oxides of nitrogen (NO_x), particulate matter and noise pollution. Especially, the greenhouse effect is much more serious in the higher part of the stratosphere, which is underlined by the fact that the UN's committee on climate change (IPCC) recommends that all measures of CO_2 from airplanes should be multiplied by a factor of 2.7. This indicates that climate impacts from airplanes are almost three times as high as other means of transportation (IPCC, 1999). An airplane trip of 12,000 km, for example Copenhagen to New York and back, contributes with 4.5 metric ton CO_2 per passenger (Nielsen, 2001). For comparison, in 2006, an average Danish citizen's total annual CO_2 emission was 10–11 metric tons, without international air travel included. A tourist consumes about 90% of the primary energy required for a holiday for transportation during the outgoing and returning journey; and a significant part (42%) of all international tourist arrivals are made by airplanes (Gössling & Peeters, 2007; Leitschuh-Fecht, 1998).

Compared with other means of transportation, air traffic is globally the least regulated (see Lassen, 2005). Airline fuel has been exempt from tax since 1944 under the rules of the International Civil Aviation Organisation (ICAO) (Lassen et al., 2006). Emissions from international flights are not included in the Kyoto protocol on greenhouse gases because of difficulties in allocating emissions between countries. Moreover, the European Commission has shown that the external costs of airplanes measured in EUR/1000-personkilometres are more than double the external costs trains (Europakommissionen, 2001, p. 116). Both in Denmark and Sweden, the building of a new airport is very much a local matter, which is not particularly governed by any coherent national plans for the future development of airports and aeromobility.

In relation to local initiatives focusing on attracting more experience travellers with airplanes, it seems problematic that the aviation industry is not subject to any international regulations, although it has serious global environmental impacts, and that environmental emissions from airplanes are not part of local environmental protection initiatives. For example, the main environmental impacts recognized by Billund airport in their annual environmental rapport are: noise, wastewater, percolation to groundwater store, disposal of liquids from the de-icer platform, waste and air pollution (traffic related, energy use, etc.). Those are all local environmental impacts. In other words, focus is solely on reduction of environmental impacts at company level, i.e. the airport itself.

With regard to the reduction of air polluting substances, the calculations do not encompass contributions from planes. Seen in the light of the global climate problems, it almost seems ironic that Billund Airport has received a diploma three times for its environmental reports (1999, 2002 and 2004). The environmental diploma is handed out by the Danish Environmental Network South (Miljønetværk Syd), a regional cooperation project between private companies and public authorities on environmental matters.

Low-Price Air Companies and Secondary Airports

The story of Stockholm Skavsta airport is an example of one of many underused secondary airports across Europe that has been brought back in use by low-cost airlines. Even though Billund airport has a different history, it is also (as in Nykøbing) the increased number of low-price departures that has been the central element in the progress of the airport during recent years. The commencement of this resurgence took place in 1997 when the air industry of Europe was deregulated (Barrett, 2004a). The pre-deregulation system of airline regulation left most secondary airports empty. Before deregulation there was little impetus to develop these airports. The high-cost base of traditional airlines made it necessary for them to concentrate on hubs for the purposes of interlining and coordination (Barrett, 2004a). These non-hub airports, which previously had been developed by various city and regional governments and in some cases, private investors, became a vital part of the success of low-cost airlines, in particular companies like Ryanair, EasyJet, etc. (Barrett, 2004a).

The case against secondary airports was their remoteness compared with hub airports and the difficulty of access from major destinations. However, travellers have transferred strongly to new airports in Europe, and it is fair to say that secondary airports have become part of the low-cost airline product (Barrett, 2004a, p. 91). As one observer notes, the low-cost carriers and secondary airports across Europe developed a "successful marriage", and as a couple they grew to like each other more as time went by (Barrett, 2004b). They both developed a low-cost corporate culture instead of the protectionist culture that was prevalent in hub airports and high-cost airlines. Some of the main elements in the low-cost culture were the fact that low-cost airline products did not require business lounges and airbridges, and needed fewer check-in desks because check-in was not slowed down by interlining, seat allocation, frequent flyer point allocation, etc. For travellers, the main attraction to the secondary airports was of course fare savings, and also the simplicity of new airports, short walking times, quicker check-in, cheaper parking costs, less confusion and better punctuality. This combination made it possible to reduce turnaround time significantly, which meant that planes spend more time in the air and less on the ground and do more rotations per day (Barrett, 2004b). It is such change patterns of aeromobility that Billund and Nyköping see as their opportunity if they can manage to create interesting places to visit for tourists.

Conclusion and Perspectives

In conclusion, we discuss some of the planning and policy implications from the two case studies. The overall theme of this article is to explore how aeromobility is used as a core element in the development of new urban strategies of experience and transformation of urban spaces. In various ways, Billund and Nyköping illustrate how (low-price) flights can be used as a tool to connect places to the global space of flow of experience tourists and to attract new settlers. In both cases, the local airport has been a key actor in the attempt to attract air companies and reshape the cities. The local spatial transformation and development as an experience destination involves a network of local, national and global forces, money, ideas and mobilities. It focuses on attracting new types of more mobile lifestyles and identities of tourists and settlers (Nielsen, 2001; Urry, 2007). This conclusion seems, however, to call for much more research in the future concerning the complex relationship between increased air travel and social and spatial reorganization.

From an environmental perspective, a number of critical questions can be raised, involving a multitude of ethically challenging issues and aspects that needs to be dealt with by planning and policy-making systems on various levels. The cases of Billund and Nyköping contain important knowledge for planners and policy-makers at several levels—at the EU level, national level and the regional/local level as well as for the links between these levels. In cases like Billund and Nyköping, we are facing a complex connection between the increasing possibilities of flying globally and local–global reorganization of urban activities and the transformation of urban spaces on various levels. This means that new air-flight destinations constantly arise, which supports more flying rather than less.

On the local/regional level, it seems problematic to base experience economy on increasing flights because of the serious environmental impacts connected to air transport. A city that wants to develop a local experience economy based on increasing aeromobility has to consider whether this is compatible with environmental responsibility and being "a green city" that lives up to its Local Agenda 21 obligations. Additionally, (aero)mobility is, as Cresswell (2006) points out, a question of power. The cases of Billund and Nyköping also illustrate a number of discursive power aspects such as the presumption that access to global flows of people and tourism will necessarily improve the economy and social situation of a city or a region. The global space of flows is not only a material way to organize international movement; it also represents a rationality of actions. For regions around Billund and Nyköping, there is a risk of just becoming another transit place with all the negative social aspects which are involved, instead of becoming a new place that tourists want to visit. The cases of Billund and Nyköping show how the norms of aeromobility are adopted by the local governments without any reflection of the negative local and global environmental and social impacts.

However, the opportunity for regionally developing new urban strategies based on aeromobility is, to a high extent, supported by the development of the air transport sector, which, as described earlier, is one of the least regulated policy areas, both at the national and the global level. The air sector is therefore based on a market-driven rationality with focus on constantly finding new destinations and increasing the number of air travellers. And as Gössling and Peeters show by analysing the dominating discourses surrounding air travel, four major air industry discourses can be identified: (1) air travel is energy efficient and accounts only for marginal emissions of CO_2; (2) air travel is economically and socially too important to be restricted; (3) fuel use is constantly minimized and new technology will solve the problem and (4) air travel is treated unfairly in comparison with other means of transport. Such discourses, and the discourses of space of flows and global city competition, make it very difficult to articulate a critical perspective on flying and new patterns of experience and play at the national, European and global levels. Therefore, this article will conclude by suggesting that a much stronger public debate of the environmental consequences of air travel is needed. Such a debate needs, in particular, to focus on the introduction of an international tax system for air travel and new institutional systems that can support a more sustainable development of air travel.

Acknowledgements

We would like to thank Erik Jensen (Roskilde University) for valuable comments on this article.

Note

1. The term "aeromobility" relates to process of air traffic as a parallel to automobility (Urry, 2000, p. 59). The term aeromobility is inspired by Høyer (2000, p. 193). Furthermore aeromobility refers in this article to both actual air trips of individual and their capacity to carry out air-based mobility (Kaufmann, 2002, p. 1). This means that to understand the production of air traffic one may not only study peoples actual movement, but also their potential to carry out different types of mobilities and in relation to this understand which mechanisms that transform/not transform potential mobility into actual mobility.

References

Adams, J. (2005) Hypermobility: A challenge to governance, in: C. Lyall & J. Tait (Eds) *New Modes of Governance: Developing an Integrated Policy Approach to Science, Technology, Risk and the Environment*, pp. 123–138 (Aldershot: Ashgate).

Barrett, S. D. (2004a) The sustainability of the Ryanair model, *International Journal of Transport Management*, 2(2), pp. 89–98.

Barrett, S. D. (2004b) How do the demands for airport services differ between full-service carriers and low-cost carriers? *Journal of Air Transport Management*, 10(1), pp. 33–39.

Bærenholdt, J. O. & Sundbo, J. (Eds) (2007) *Oplevelsesøkonomi: Produktion, forbrug, kultur* (København: Samfundslitteratur).

Billund Airport (2007) *Statistics* (Billund: Billund Airport).

Castells, M. (1996) *The Rise of the Network Society. The Information Age: Economy, Society and Culture*, Vol. 1 (Oxford: Blackwell).

Cresswell, T. (2006) *On the Move. Mobility in the Modern Western World* (London: Routledge).

ErhvervsBladet (2007) Billund oplever boom i passagerer, *ErhvervsBladet*, December 4, pp. 4–5. Available at http://www.erhvervsbladet.dk/article/20071204/news06/71203006 (accessed 29 January 2008).

Europakommissionen (2001) *Hvidbog. Den europæiske transportpolitik frem til 2010—Det svære valg* (Bruxelles: Kommissionen for de Europæiske Fællesskaber).

European Low Fares Airline Association (2007) *Social Benefits of Low Fares Airlines in Europe*, Available at http://www.elfaa.com/documents/Social_Benefits_of_LFAs_in_Europe_(York)_211107.pdf (accessed 29 January 2008).

Frändberg, L. (1998) *Distance Matters: An Inquiry into the Relation Between Transport and Environmental Sustainability in Tourism*, Humanekologiska skrifter 15 (Göteborg: Göteborgs Universitet).

Gössling, S. & Peeters, P. (2007) "It does not harm the environment!" An analysis of industry discourses on tourism, air travel and the environment, *Journal of Sustainable Tourism*, 15(4), pp. 402–417.

Graham, B. (1995) *Geography and Air Transport* (Chichester: John Wiley).

Groth, N. B., Lang, T., Johansson, M., Kanninen, V., Anderberg, S. & Cornett, A. P. (Eds) (2005) *Restructuring of Medium Sized Cities —Lessons from the Baltic Sea Region* (Copenhagen: Danish Centre for Forest, Landscape and Planning).

Hall, P. (2000) Creative cities and economic development, *Urban Studies*, 37(4), pp. 639–649.

Halldin, K. & Hultman, M. (2005) *Destination Development — Perspectives on Co-operation and Attitudes in Networks: A Case Study of Nyköping* (Lund: Lund University).

Høyer, K. G. (2000) Sustainable mobility—The concept and its implications, PhD dissertation, Institute of Environment, Technology and Society, Roskilde University, Roskilde.

Høyer, K. G. & Næss, P. (2001) Conference tourism: A problem for the environment, as well as for research? *Journal of Sustainable Tourism*, 9(6), pp. 541–570.

IPCC (1999) *Aviation and the Global Atmosphere,* Special Report of IPCC Working Groups I and II, Intergovernmental Panel on Climate Change, Cambridge: Cambridge University Press.

Kaufmann, V. (2002) *Re-thinking Mobility. Contemporary Sociology* (Hampshire: Ashgate Publishing Limited).

Kunzmann, K. (2004) Culture, creativity and spatial planning, *Town Planning Review*, 75(4), pp. 383–404.

Lassen, C. (2005) Den mobiliserede vidensarbejder: En analyse af internationale arbejdsrejsers sociologi, PhD dissertation, Institut for samfundsudvikling, Aalborg Universitet, Aalborg

Lassen, C., Laugen, B. & Næss, P. (2006) Virtual mobility and organizational reality: An examination of mobility needs in knowledge organisations, *Transportation Research Part D*, 11(6), pp. 459–463.

Leitschuh-Fecht, H. (1998) *Tourism and Sustainable Developmen*, Report on the 7th Meeting of the Commission on Sustainable Development (CSD), Bonn, November.

Lorentzen, A. (2009) City roles under transformation: An evolutionary approach, *European Planning Studies*, 17(6), pp. 925–941.

Marling, G., Jensen, O. B. & Kiib, H. (2009) Design and brand in the experience city—Hybrid cultural projects and performative urban spaces, *European Planning Studies*, 17(6), pp. 863–885.

Metz, T. (2002) *FUN. Leisure and Landscape* (Rotterdam: NAi Publishers).

Nielsen, S. K. (2001) Air travel, life-style, energy use and environmental impact, PhD dissertation, Technical University of Denmark, Copenhagen.

Nyköping Kommun (2007) *Besöksnäringsstrategi 2008* (Nyköping: Nyköping Kommun).

Pascoe, D. (2001) *Airspaces* (London: Reaktion).

Pine, B. J. & Gilmore, J. H. (1999) *The Experience Economy: Work Is Theatre and Every Business a Stage* (Boston: Harvard Business School Press).

Shaw, G. & Williams, A. M. (2004) *Tourism and Tourism Spaces* (London: Sage).

Smidt-Jensen, S. (2005) *City Branding: Lessons from Medium Sized Cities in the Baltic Sea Region* (Copenhagen: Danish Centre for Forest, Landscape and Planning).

Therkildsen, H. P., Hansen, C. J. & Lorentzen, A. (2009) The experience economy and the transformation of urban governance and planning, *European Planning Studies*, 17(6), pp. 925–941.

Urry, J. (1990) *The Tourist Gaze* (London: Sage).

Urry, J. (1995) *Consuming Places* (London: Routledge).

Urry, J. (2000) *Sociology Beyond Societies: Mobilities for the Twenty-First Century* (London: Routledge).

Urry, J. (2007) *Mobilities* (Cambridge: Polity Press).

Whitelegg, J. (1997) *Critical Mass: Transport, Environment and Society in the Twenty-First Century* (London: Pluto).

Experiential Strategies for the Survival of Small Cities in Europe

PETER ALLINGHAM

ABSTRACT   *The aim of this article is to analyse, discuss and evaluate different methods of branding applied in experiential strategies for the survival of small cities in Europe. After the introduction that refers to the advent of the experience economy in the post-Fordist era, the article introduces various branding methods applied in experiential strategies. Then follows an analysis of how those branding methods are applied in experiential strategies for the development and survival of two small cities in Germany, Dresden and Wolfsburg, in which car production and city development have been combined. The article concludes with an evaluation of the branding methods, which includes considerations of whether they can be used as models of survival for other small European cities. The evaluation refers to recent views on the question of representation and authenticity, and the role of cultural heritage in experiential strategies.*

Introduction

We do not build architecture of assertion or representation; we occasionally call our work neuronal architecture or process architecture. (Gunter Henn in Ahrens, 2001, p. 204)[1]

Since the collapse of the "cold war world" in 1989, symbolized by the fall of the Berlin wall, a redrawing of both geographic and mental boundaries between nations and regions in Europe has taken place. In a climate of post-Fordist economy characterized by accelerated turnover time, intensified competition and a diversion of consumer attention towards brands, images and experiences, many cities in Europe have had to adopt new strategies to enter sustainable economic development.

In 1999, the economists B. Joseph Pine II and James H. Gilmore summarized the economic situation in an influential book with the title *The Experience Economy. Work is Theatre & Every Business a Stage*. It introduced experience economy as the most recent step in a continuous economic development, in which businesses, in order to survive, instead of servicing consumers should engage them in experiences by "wrap[ping] experiences around their existing goods and services to differentiate their offerings" (Pine & Gilmore, 1999, p. 15). This advice has, of course, led to several questions and problems. What are experiences? And how is the "wrapping" done?

According to Pine and Gilmore experiences are inherently personal. "They actually occur within any individual who has been engaged on an emotional, physical, intellectual, or even spiritual level." (Pine & Gilmore, 1999, p. 12). Another more psychological definition states that "an experience originates from an individual's emotional and cognitive processing of the sensual impressions (stimuli) that the organism receives from a world of objects. Experiences presuppose active contributions from the person who experiences" (Jantzen & Vetner, 2006, p. 240).

Important points to be extracted from those definitions are, first, that consumers of goods and receivers of market communication have become even more pivotal in strategic thinking and planning than ever. Furthermore, communication theory has become a key discipline in strategic design of experiences, and branding, i.e. the practice of attaching a name or a mark to an object in order to distinguish it from similar objects, has become one of the most important strategic tools in experiential design. Although commercial branding has been practised since the end of the nineteenth century, its importance as a means for creating identity, image and experiences has been growing (see Olins, 1989; Kunde, 1997; Mollerup, 1997). Secondly, the psychological definition above draws attention to the importance of understanding the exchanges between mental processes and the object world. These exchanges imply representation and aesthetic mediation. Therefore, aesthetics understood as formed organization of a material is nodal in the creation of experiences because aesthetics evoke psycho-physical response and semiotic relevance (Research Group MAERKK, Aalborg University, see references). This certainly draws attention to the potentials of urban planning, architecture and design as branding tools in experiential strategies to create identities for people, communities and places, among them cities.

Aim and Methodology

It is central to this article to analyse and evaluate how branding methods are applied in experiential strategies for the survival of small cities in Europe. First, various branding methods applied in experiential strategies will be introduced. Secondly, experiential strategies applied in two small German cities will be analysed. The first city is Dresden in Saxony with approximately 500,000 inhabitants. The other is Wolfsburg in Lower Saxony with approximately 120,000 inhabitants. These cities have been through different types of crisis, among them the phasing out of old-type industrial production, shrinking numbers of inhabitants and in the case of Dresden, a conversion to a western liberal market economy. However, in various ways both cities have adapted to an ongoing development from an industrial culture to a culture of information, knowledge and experience. In both cities, Volkswagen produces and markets cars and takes part in urban development by applying different branding methods in the realization of experiential strategies.

The approach of the article has been to combine humanistic theories of communication and media with elements from discourse analysis and material gathered in field studies and observation studies, using methods of naturalistic inquiry (see Lincoln & Guba, 1985), during visits in Dresden and Wolfsburg through 2006–2008. Material has also been compiled through comprehensive photographing, gathering and reading of printed material like brochures, books, and through accidental interviews and a meeting with representatives of Wolfsburg AG. The material provided through observation is analysed like a "text" with the point of departure in a methodological set-up that draws on, first, Roman Jakobson's work on communication and representation (Jakobson, 1971) and Charles S. Peirce's semiotic typology of signs (Peirce, 1931–1958) semiotic typology of signs. Secondly, the concept of "blanks" introduced in reception theory will also be applied (Iser, 1974; Eco, 1979, 1990). "Blanks" refer to gaps or indeterminate spaces in meaning structures into which receivers of communication during interpretation insert or fill in their own often aberrant or even subversive meanings. Finally, Michel Foucault's concept of "dispositive" is applied. This concept originates in his analyses of discourse and power. It refers to the silent, non-physical agency of power that stretches through and permeates networks of knowledge and rationality (Foucault, 1980, p. 194). In the present analyses, this concept refers to power structures like governance and control disposed through branding methods as implementations of experience economic strategies. Here, governance refers to the procedures, practices and incentives in both the public and private sectors by which decisions are taken. Specifically, good governance in public–private partnerships (PPPs) refers to those procedures, practices and incentives which are associated with the delivery and preparation of infrastructure projects (Hamilton, 2007). The analyses will focus on where the applied branding methods, representing dispositive rationality and knowledge, become unstable and invite critical response involving the memory and/or knowledge of receivers. In this way a critical platform can be obtained from which the experiential strategies as means for the survival of cities can be evaluated.

Branding

In general, since the early 1990s branding has developed into one of the most preferred means by which visibility and identity can be assigned to everything from commercial products and corporations to countries, regions, cities, neighbourhoods, kindergartens and persons. Especially the development of corporate branding has been significant in connection with cities and regions. By corporate branding we understand a type of branding that comprises all levels and layers in a company and not only a certain product.

In many cases, like in the case of the Body Shop, corporate branding has meant, on the one hand, that companies have become more "transparent". From the outside, customers and stakeholders have continuously been able to follow the actions, behaviour and moral standards of a company. On the other hand, corporate branding has been seen as a means of strategic market communication through which a company would be able to control its outside world. Due to the collapse of a rigid distinction between internal and external communication, employees and external stakeholders have increasingly been regarded on an almost equal level as both recipients and co-producers of the products and services of a company. This rearrangement of the traditional roles of senders and receivers in corporate communication has promoted a state of communication where companies in principle would be able to produce and so, apparently, to a certain extent control their own external reality in a comprehensive "bubble of communication". However, they are also constantly in danger of losing touch with the reality outside the bubble (see, e.g. Christensen, 1994; Christensen & Cheney, 2000).

So far, this type of branding, among certain observers called "Branding 1.0" or "Old School Branding", has been employed through the 1990s and into the present decade with disputed success by a number of regions, cities and organizations aiming to become more visible in the local and global competition for attention (see Stigel & Frimann, 2006). One of the reasons why the corporate branding method has not been quite so successful is that private and public organizations are founded on different structures of decision making. Public organizations like municipalities are, unlike many private organizations, founded on democratic principles where disagreement is fundamental. Therefore, since corporate decision making in most cases is not democratic, Branding 1.0 is difficult to implement in public organizations encompassing cities or regions.

However, a more successful branding technique may be in the pipeline. A growing demand for individualized and sophisticated experiences among consumers increasingly decides how producers of goods and services should organize their businesses strategically. And it seems that those who can find out how to cater as closely as possible to the individual consumer's fantasies and wishes will stand a better chance to survive in the long term. Therefore, the making of creative and interactive spaces and frames, in which various types of experientially oriented audiences can perform and unfold their projects, seems to be an important element in a new type of branding. This type of branding has been called "Branding 2.0", "The Learning Brand" or "Network Branding" (Buhl, 2005; Grant, 2006). Today, a number of three-dimensional frames and types have emerged, in which individual experiential projects may proliferate in cooperation with input from public or private operators, e.g. brandscapes and brandhubs. Brandscapes are three-dimensional designed brand settings (Riewoldt, 2002, p. 7), and brandhubs are "comprehensive urban mixed-use environments developed by brand-name corporations in partnership with host city authorities. Utilized as branding instruments and designed by signature architects, they aim to mediate corporate identities to a broad audience in an experiential 'public' space" (Höger, 2004, p. 125).

In the following paragraphs two examples of branding methods are analysed, representing Branding 1.0 and 2.0 used in experiential strategies in order to develop two cities. The analyses will focus on how aesthetics, memorial qualities and historical legacy have been applied. The question is whether the examples of Dresden and Wolfsburg present possible models of survival that might be adopted by other mid-sized and small cities in Europe.

Dresden: Brief Introduction

Dresden, lying on both banks of the river Elbe in Saxony, has a long history as capital and royal residence for the Electors and Kings of Saxony. For centuries they furnished the city with cultural and artistic splendor, such as cathedrals and palaces—many of them in the Baroque style, e.g. the Zwinger Castle, Frauenkirche and many others. However, the controversial bombing of Dresden in World War II and 40 years in the Soviet bloc state of the German Democratic Republic (henceforth GDR), followed by contemporary city development, has significantly changed the face of the city.

Die Gläserne Manufaktur

Since the German reunification in 1990, Dresden has emerged as a cultural, political and economic centre in the eastern part of Germany. In spite of unemployment rates between

15% and 20% and inflation after the adoption of the Euro currency in 2002, considerable restoration works have been carried out.

In an effort to boost economy and relieve unemployment, VW has built an assembly plant for their luxury limousine, the Phaeton, in the centre of Dresden. The name of the plant is Die Gläserne Manufaktur. It was created and designed by Gunter Henn, Chief Architect of Henn Architekten, who is also behind AutoStadt in Wolfsburg.

The name of the plant refers to the fact that the factory building is transparent, as most of it is a huge glass structure. This makes it possible to watch the assembling of the cars from both inside and outside the building. The factory lies close to Dresden's historic city centre in a corner of the famous Baroque park Grosser Garten, with a history of being a royal hunting ground 300 years ago.

With Die Gläserne Manufaktur, the concept of a city and region of experience has been given quite a unique design. The main point is that the act of buying a hand-built VW Phaeton has been scripted as an experiential sequence that carries the characteristics of an initiation into the world of culture. The act of buying includes, apart from customizing activities, an invitation from VW to spend a week at Dresden. During the stay the customer is enrolled in a cultural programme that includes visits to a number of cultural and historical attractions in Dresden, a night at the Semper Opera, a tour of the Zwinger, an afternoon at Schlosspark Pillnitz, etc.

The opening of Die Gläserne Manufaktur in December 2001 by the then German Chancellor, Gerhard Schröder, did not occur without a previous long heated debate between protesting citizens, VW and public authorities in Dresden. When it became public that VW planned to build the plant in the heart of Dresden, more than 17,000 citizens signed a petition against it. The protesters held that heavy traffic with lorries loaded with car parts and perhaps smoking chimneys would pollute the air and spoil the ambience of the city centre, and they opted for the new plant to be placed in one of the many deserted factories in the outlying areas of Dresden. However, the protests were abated when it was agreed that the car parts were to be carried to the factory by special Cargo Trams on the city tram rails from depots in the suburbs. And after VW had presented a plan for an integrated project that respected the history, tradition and environment, the Grosser Garten location was finally selected.

According to the German architectural critic Andreas Ruby the urban planning behaviour of VW has been very sensible towards the spatial and historical urban context of Dresden (Ruby, 2000a). First, the angular volume of Die Gläserne Manufaktur continues the city space instead of blocking it. Secondly, the architecture repeats the open, green built areas of the adjoining neighbourhoods from the GDR epoch. Thirdly, to the south Die Gläserne Manufaktur relates to the Grosser Garten by also being embedded in a park-like landscape with a lake. Historically, Die Gläserne Manufaktur continues a local tradition of exhibitions that has been going on at the site for more than a hundred years. Until the end of the nineteenth century, the "Crystal Palace" exhibition building of Dresden stood here.

Two further adaptations to the surroundings can be observed. First, like the hothouses in the neighbouring botanical gardens, Die Gläserne Manufaktur is also an artificial "biotope" for something that would not survive outside it, viz. the "Manufaktur". This transparent "theatre of work", where workers in white overalls and white gloves standing on parquet floors can be seen assembling the cars, refers to pre-industrial cultural traditions in the region of the manufacture of, for example, porcelain in the close-by city of Meissen.

Secondly, as pointed out in a promotional DVD film about Die Gläserne Manufaktur, the transparency of the building refers to the anatomical model Der gläserne Mensch exhibited at the Deutsches Hygiene-Museum only a few hundred metres from Die Gläserne Manufaktur towards the city centre (Gingco.net Werbeagentur, 2005). In this model, skin and muscles have been made transparent in order to make inner organs visible. Likewise, because of the transparency, a symbolical suspension of the limit between factory and city has taken place.

Baroque hyper-adaptation
In an interview about Die Gläserne Manufaktur, Chief Corporate Architect Gunter Henn has stated:

> Given the valency of the building and its special situation on the fringe of Dresden's Altstadt not far from the Semper Oper, Frauenkirche and Zwinger, the corporate culture is not only made visible; it must also assert itself in the space of value and meaning of the city of Dresden. (Ahrens, 2001, p. 204)

Gunter Henn's description of a reciprocal appropriation of factory and city implies an experiential strategy in accordance with the "Old School" corporate branding model, in which a corporation tries to control and produce its environment by incorporating and adapting it to its corporate culture and identity. In the case of Die Gläserne Manufaktur it is a question, however, of how equal the reciprocal adaptation between factory and city has been. Certainly, the Baroque attractions of Dresden serve as a cultural backdrop and provider of identity to the manufacturing process and the VW Phaeton limousines. But the question seems to be whether Die Gläserne Manufaktur has not hyper-adapted itself out of a sustainable identity. Not only is the factory building transparent, it is a "theatre" staging a play. Furthermore, the factory has been bended, twisted and groomed to adapt to the environment and the emotions of protesters, which is generally considered a sign of loss of identity and culture (Hatch & Schultz, 2004, p. 392). The factory turns out an anonymous wonder car whose name, the Phaeton, refers to an educated culture and appeals to affluent segments with a romantic or nostalgic sentiment.

Arguably, the coveted corporate identity equation says that the VW Phaeton and Die Gläserne Manufaktur are authentic pieces of cultural, Baroque Dresden. Therefore, an owner of a Phaeton will, logically, take over and live these cultural values. However, there is a problem with the equation. The problem is not that it is unimaginable that physical environment and city space can be part of a corporate culture. The problem in the case of Dresden is that the cultural identity promoting values rely on what seems to be a reconstruction, a simulacrum.

This appears in an article by the architectural critic Andreas Ruby with the title "Las Vegas an der Elbe" (Ruby, 2000b). Here, one of the major points is that recreating a city in the picture of a selected past cultural epoch, the Baroque, in order to establish a clear identity, produces the opposite effect. Instead of acknowledging that Dresden has a multi-faceted historical background that includes a World War II and the GDR epoch, Ruby concludes that the city will become a simulacrum, a klichée for the tourists, ultimately comparable to Las Vegas.

Seen from the viewpoint of VW, it has been important that the setting and staging of the manufacture of the Phaeton should not take place in any ordinary experiential theme park.

Because, what the customer should see through the aesthetic mirror prism of the VW Phaeton is a reflection of himself as an integral part of the authentic ambience of a historical and cultural city like Dresden. This is similar to the anticipation created in relation to the experiential Phaeton Konfigurator-computer at the experiential basement in Die Gläserne Manufaktur. On this computer, the visitor can design his own model Phaeton and project it and himself into a number of selected cultural scenarios from the city of Dresden, and anywhere else that suits his dreams. In Figure 1, the author of this article can be seen standing beside his Konfigurator generated Phaeton model in front of Die Gläserne Manufaktur.

However, in spite of being a physical fact, the Phaeton car is, like Die Gläserne Manufaktur, a phantom. In both cases their credibility as strategic implementations in extension of an experiential economic dispositive rationale is eroded in several ways. First, the transparency is ambiguous. True enough, transparency promotes aesthetic quality, artifice and presentation. It is supposed to promote free access, openness and creativity so that everybody can see what is true—namely, and as Gunter Henn has phrased it, that Die Gläserne Manufaktur could be "Zwinger of the 21st century" (Rauterberg, 2001). However, aesthetics and presentation are disguises. Why else should the discretely operating "Putzfrauen" hovering in the wings of the showrooms at Die Gläserne Manufaktur go into action whenever a team of visitors have left a showroom? When the visitors have left, they come out to remove all fingerprints from the car models on display. The fact is that indexing fingerprints represent, as they refer to use and a history of use, i.e. history, which means that they disturb, blur and intrude into aesthetic presentation and the possible enchanted experiences that originate from it. Ultimately, the corporate branding project seems to appear as a project of ideological aestheticization, in which profane Fordist assembly-line production has been "wrapped" as Manufaktur and cars as cultural education and/or art.

Figure 1. The author's "dream" came true in print
Source: Konfigurator photo taken by P. Allingham at Die Gläserne Manufaktur, Dresden, April 2007.

Wolfsburg 1938–2008: A Brief History

The development of Wolfsburg is closely connected to the VW car brand, in fact Wolfsburg owes its existence to this car brand, which is why we have to recapitulate in brief terms the history of the car and the city.

The foundation stone of the Volkswagenwerk was laid at the village of Fallersleben on 25 May 1938 with full Nazi pomp and circumstance. Shortly after, on 1 July 1938, a city called "Stadt des KdF-Wagens", the Wolfsburg of today, was founded as a city for workers and employees at the Volkswagenwerk. At the Volkswagenwerk, Adolf Hitler, who had been inspired by the successful concept of the "Volksempfänger" radio, wanted to realize his dream of a car within economic reach of the average worker. Ferdinand Porsche had developed the car, KdF-Wagen, after the war known as VW-Käfer. It was sold on a saving stamp scheme through "Kraft durch Freude", which was the recreational community of the national socialist party.

During the World War II the factory was bombed. Despite being severely damaged, car production was resumed in 1945, headed by the British Major Ivan Hirst. The city was renamed Wolfsburg, after the castle close-by, and a plan for the housing of 35,000 inhabitants was introduced. In 1948, the Volkswagenwerk was given over to German hands. During the 1950s and into the 1970s, the production of an expanding range of VW car brands became a major success.

Until the reunion of East and West Germany in 1989, the city of Wolfsburg was generally considered to be an unattractive industrial dead-end close to the zonal boundary on the GDR. Recessions in the 1990s after the reunion and a drastic fall in the demand for VW cars during 1992–1993 hit hard on both the VW-group and the Wolfsburg region. During the increasing welfare in Europe, VW developed into a multinational company with production units in several countries around the world. This development created a need for a new dynamic company image as provider of high quality products associated with a modern lifestyle and feeling of life. This led, towards the end of the 1990s, to a restructuring of both corporation and city that aimed at freeing the city from its image as obsolete, stagnant and mono-industrial. Instead, a new attractive dynamic identity of diversity, creativity and experience was launched.

Wolfsburg AG

In 1999, Wolfsburg AG was founded as a limited PPP between VAG and the city of Wolfsburg. Wolfsburg AG is described as "the powerhouse behind the successful implementation of AutoVision in Wolfsburg [with] the creation and development of company networks and clusters" (Wolfsburg AG (Hrsg.), 2007 p. 8). AutoVision is a scheme developed by VAG during the 1990s on the basis of thorough research into future socio-cultural trends (Pohl, 2005). It was aimed at cutting unemployment by 50%, developing a new sustainable economic structure and transforming Wolfsburg into a dynamic and economically strong region offering a high quality of life to both residents and visitors. The intention has been to create new jobs, especially service jobs, facilitate company start-ups, invigorate the location and make it more attractive. These activities have been organized in Wolfsburg AG in five business segments: an InnovationsCampus, a MobilityBusiness, a LeisureBusiness, a HealthProject and a PersonnelServiceAgency (Figure 2).

Each segment aims at contributing to the development through a number of initiatives. The most interesting and spectacular, in this connection, is the LeisureBusiness with the

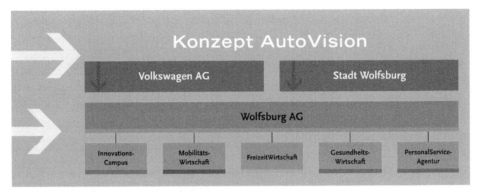

Figure 2. Wolfsburg AG structure
Source: Courtesy of Wolfsburg AG.

ExperienceWorld. It aims at developing a number of high-standard entertaining and experiential attractions for tourists and citizens.

However, it is important to realize that the AutoVision merging of the aspects of labour market, innovation, urban development, identity and image has its roots in the VAG's business culture and strategic visions. In fact, the AutoVision concept was given by the VW-group to its home city as a present on its 60th birthday in 1998.

ErlebnisWelt
On the homepage of Wolfsburg AG it is pointed out that the ErlebnisWelt concept, with the slogan "To move is to live" stands for fun, sport, excitement and entertainment in six differ-ent key areas. These areas are: "Sports and Recreation"; "Discovery and Entertainment"; "Shopping Experience and Variety"; "Art, Culture and Lifestyle"; "Tradition and Moder-nity" and finally "Fun and Fantasy". The ambition behind all the entertainment is to create new jobs in the service sector, to develop a cluster of businesses involved in tourism and to encourage the emergence of new leisure concepts that will spark international interest.

In order to accomplish this, the ErlebnisWelt concept has been realized by means of an "emotional mapping" of the city of Wolfsburg and surrounding areas (Wolfsburg AG (Hrsg.), 2007, p. 13). The map in Figure 3 shows a distribution of the six key areas of experience in and around Wolfsburg.

As it appears in Figure 3, the key experiential areas of Wolfsburg are located according to an axial distribution that radiates from the hub of the car theme park, AutoStadt. However, although it has no immediate reference to the experiential planning apart from tourist boat trips, a first axis to be noticed is the Mittellandkanal. It severs both the city and the surrounding landscape into two halves, referring to an infrastructural div-ision between industry and habitation from the past. Then, at least two imaginary experi-ential axes can be distinguished, a strong and a weak one, connecting the emotional key areas. The strong axis stretches vertically between the fields of "Tradition und Moderne" and "Kunst, Kultur und Lebensgefühl", from the Castle of Wolfsburg through AutoStadt, across the Mittellandkanal, past the PhaenoScienceCenter, along Porschestrasse to the Museum of Art. Stretching your imagination, another possible axis can be distinguished running from AutoStadt through the Allerpark to the Fun and Fantasy area across the Mittellandkanal.

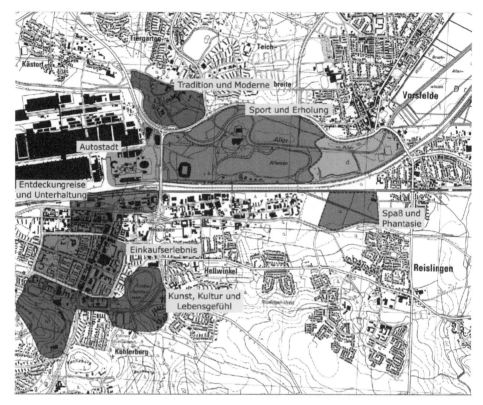

Figure 3. Key areas in the ErlebnisWelt of Wolfsburg
Source: Courtesy of Wolfsburg AG.

This radiating structure certainly demonstrates and refers to the key role of AutoStadt as experiential hub. According to Pohl (2005, p. 644) from the Geographic Institute at the University of Bonn, Autostadt is the primary component and corner stone upon which the structure of the "ErlebnisWelt" has been founded.

As it appears from this statement, the overall key in Wolfsburg's transition from industry to experience is closely connected to VAG and its economic power, enterprises and initiatives, with AutoStadt as a unifying hub. The overall strategic dynamic experiential scheme appears from this organization and from the reference in various types of promotion to "movement" and "mobility", e.g. in the slogan "Menschen, Autos und was sie bewegt". The idea is simply to attract customers and guests to AutoStadt, "charge" them with experiences and creativity and finally radiate them along the spokes of the experiential matrix into the experiential zones of Wolfsburg, some of which are still under construction.

In the following paragraph an analysis is presented of how AutoStadt, "hub" in the experiential "wheel", fulfils the aims of the AutoVision concept.

AutoStadt: aesthetics as experiential catalyst
The 25 hectare AutoStadt at Wolfsburg opened on 1 June 2000 in connection with the World Exhibition in Hanover, the EXPO2000 (Figure 4). It claims to be the only existing

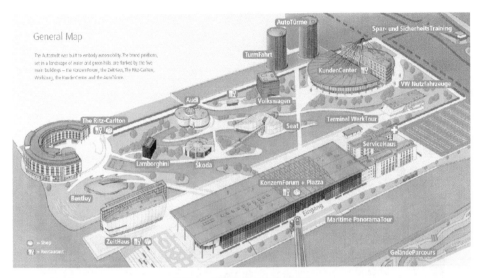

Figure 4. Map of AutoStadt
Source: AutoStadt GmbH.

tourist-oriented car theme park. It calls itself a futuristic theme park on the subject of "automotive mobility" (Wolfsburg Marketing GmbH (Hrsg.), 2006, p. 11).

Apart from being VAG's communication and service platform and physical interface to the surrounding world, it serves at least three purposes. First, it is a piece of strategic marketing that takes care of the branding activities of VAG. Secondly, it serves as the final stage in a sales process where the cars are handed over to the customers. And thirdly, it is the nodal part, the power centre, of the urban and regional development of the city of Wolfsburg.

Apart from those who work there, people go to AutoStadt as visitors or as customers. Every day round the year approximately 600 customers choose to collect their new VW model at the KundenCenter at AutoStadt (Eckhardt, 2002, p. 6).

Until you collect your car it is stocked in one of the two transparent Autotürme that tower above the AutoStadt as a landmark of one of the biggest producers of automobiles in Europe. The two towers contain 800 cars (Figure 5).

One of the first objects to meet the eye of anyone arriving at AutoStadt by car is a full stop traffic sign with the inscription "Smile". This symbolic index anticipates what the slogan "Menschen, Autos und was sie bewegt" also expresses, viz. that "bewegung", emotion, pleasure, is the crux of the attraction. And true, once inside the "Stadt" there is not much physical movement. Except for strolling visitors, AutoStadt appears as a frozen spatial aggregation of industrial cultural past, present and future symbolized by, for example, the striking contrast between the chimneys of the gigantic industrial power plant and the transparent communicative Autotürme that store and display the cars that are soon to be handed over to their new owners.

Most guests and customers enter AutoStadt through the KonzernForum. Many visitors then stroll to the ZeitHaus. This building contains an experientially oriented presentation of the history of mobility through a selection of historical "Milestones" in the development of the car. As this building is a key to the understanding of the experiential techniques generally applied at AutoStadt, we shall analyse its architecture and aesthetics.

Figure 5. AutoTürme, Autostadt
Source: Photo P. Allingham, April 2007.

The "ZeitHaus" is a three partite complex consisting of two separate bodies connected by a glass-covered staircase room (Figure 6). The northern part is a transparent square-windowed glass "rack" containing a collection of historical cars on four floors. The southern part is an organically shaped, closed concrete compartment.

In the transparent pixel-square-windowed building, the historic car models are presented like miniatures in a type case, separately and according to a chronological rationale, so that each car represents a stage in the technical development of the automobile. This building, then, is supposed to represent digital memory and remembrance. The closed concrete body contains various dramatizations of the automobile in social and historical contexts. This part of the complex stands for the associative, the emotional or analogous memory. The contrasts between the analogous and the digital, the square and the organic and the transparent and the opaque support a final symbolic reference of the complex to a brain with a right and a left side. In extension of this symbolism, the balconies and the staircases between the two halves make out the neuronal net that connects the two sides. It seems that the idea of the Henn architects has been that the visitors should shuttle between the two halves of the complex like impulses in a nerve net following up on impressions and ideas, and so building up their individual experiences (Henn Architekten Ingenieure, 2000; initial vignette).

The ZeitHaus is the key in the experiential matrix of AutoStadt. Through the architectural metaphor of a brain and through the aesthetics of various design stories, the individual visitor is cast as an impulse moving in a neuronal network between sensual impressions and memory or knowledge. You "sample" the aesthetics of the "Milestones" in the digital part of the building and see some of them in selected social or cultural contexts in the analogous part of the building. What has not been dramatized you must fill in yourself.

Figure 6. ZeitHaus
Source: Photo P. Allingham, January 2008.

This initiates the coveted and enthusiastic continuity or "Bewegung": neuronal impulses becoming experiences personified in the guest, hopefully driving away in his customized VW cocoon—or, alternatively, moving downtown Wolfsburg, flinging into the multiple pleasures offered in the experiential areas.

However, this matrix of the creative brain seems to cut two ways. Its focus on joyous expectation ("VorFreude") and on spotless aesthetics creates certain blanks. Although it is true that ZeitHaus is not a museum, it is hard to hold back your historical knowledge of VW related above. You have to look hard to find traces of this concise piece of history anywhere at AutoStadt or at Wolfsburg. For example, you only get very few historical facts behind the first VW car, the famous Beetle or "Käfer", apart from a reconstructed prototype on display. The AutoMuseum Volkswagen downtown Wolfsburg mostly exhibits cars.

The mere fact that AutoStadt, with all its aesthetic strategies, has as a backdrop one of the biggest monumental industrial complexes in the world constantly refers to, and invokes, a certain historical knowledge or memory, the KdF background, the production of war material during World War II and after the war, the production of endless numbers of "Käfers".

Perhaps caught in this disposition, the guest moves on to the MarkenPavillons. There is (except for the Bugatti brand) one for each of the car brands that Volkswagen AG manages, Bentley, Lamborghini, Skoda, Seat, Audi and VW. A further pavilion contains the Nutzfarzeuge.

One remarkable thing about AutoStadt is that you hardly see any logos or other loud visual brand communications. The point is that the brands are embedded in, and speak through, the elegant design of the interior and exterior of each pavilion. A major aesthetic means is clearly storytelling, choreographed by the layout of each pavilions.

One example is the Audi Pavilion. Seen from outside, the shape of the truncated and reciprocally mounted cones of the Audi pavilion only has a slight resemblance of the four-ring logo of the Audi brand. In the pavilion the guests enter a story of the life of an Audi prototype designer Max and his wife Klara and move through their studio and stylish middle-class home. Max and Klara are interactive dolls with TV sets as heads, and "they" start talking to you at sight. After a bedroom where a sleeping person's dreams of Audi racing cars are projected on a screen, you leave the family and the building by going down a spiralling ramp around a central screen-covered column, showing aspects of an active life. The ramp takes you to a sparkling new model Audi.

The choreographic design of this pavilion seems to be the merging of the rings of the Audi brand with the contemporary elegance of an upper-middle-class lifestyle. This condensation of corporate history and contemporary everyday life seems aimed at transferring or impressing the Audi identity on the guest. The guest is invited to undertake the final analysis walking down the creative ramp, so that the finishing Audi sums up the whole design experience. This layout was found in several pavilions.

AutoStadt is apparently not an attempt to build one strong unanimous and unified corporate brand identity as much as it is an attempt to make the individual guests become vitalized metaphors of their own creative vision—projected by themselves, but framed, catalysed and facilitated by VW. Guest should move according to their own (neuronal) impulses and accumulating experiences, and not formally become or be a part or member of any "Bewegung". What brand you belong to is a private individual matter. Therefore, the network-matrix hybrid of "Stadt" and "park" seems the perfect compound frame for what you may call individualized experiential communities, where "you go to your brand and I go to mine".

Brand escape

Given the above, it seems clear that if Wolfsburg AG is the motor of AutoVision and AutoStadt is the brandhub in a wheel of experiential strategy that intends to reach into the city of Wolfsburg and its outlying areas, then it is obvious that the "petrol" that moves this marketing machine is aesthetics. As stated above, AutoStadt is altogether Wolfsburg's main attraction appealing to at least two audiences. The first is a national audience of buyers of VW cars that come to pick up newly bought cars, and save money by doing so. The second is an international audience of tourists, car enthusiasts, etc., who go there to get the experience. This audience is moved by the need and desire of its single individual members for emotional pleasures ("Freude") and intellectual, creative discovery. A third audience also exists. It should certainly not be underestimated that 25% of visitors at AutoStadt are employees at the VW and their families, an interesting fact that confirms the auto-communicative success of AutoStadt and also its quality as decisive element in the heterogenous dispositive network of Wolfsburg AG (see Eckhardt, 2002, p. 6).

As it appears, many experiences are initiated through the sensual appeal of the aesthetics, for example, of designed artefacts or the architectural layout of buildings and landscape. At the ZeitHaus there are no glass, cordons or barriers between the guest and the items on display, only a discreet request not to touch and leave fingerprints on the vintage cars on display. It means that you can invest almost all your senses in your experiences, e.g. lean over a vintage car and smell the leather upholstery. Direct sensual stimulus, often hands-on, active testing seems to be a basic ingredient in the experiential matrix.

Due to the characteristics of being an aesthetic frame catalysing and stimulating the creative projection of visitors, Autostadt is an example of a brandhub in accordance with the Branding 2.0 model. However, this "Freude-durch-Auto-Mobility" concept is not without flaws. Arguably, AutoStadt is grounded on the endogenous industrial legacy of car production. But the concrete history of the legacy, related briefly above, is conspicuously subdued. It is as if AutoStadt tries to evade the fact that most guests already know or remember at least some of the history of Adolf Hitler's Volkswagen and expect some kind of presentation or appropriation of it. Therefore, the effect is that the unreflected past, as a blank for interpretation, makes itself felt exactly because it has been almost evaded. It seems that the balance between, on the one hand, history and historical knowledge and, on the other hand, sensual experiences prompted by aesthetics has tipped so much in favour of aesthetics that aesthetics has almost become anaesthetic. Olafur Eliasson's Scent Tunnel is an example (Figure 7). The olfactory experience, i.e. the sense-appealing experience is breathtaking, but lacks relational relevance. Instead of place-bound historical facts, we are presented with, and told, stories, be it of "Milestones" in the ZeitHaus, or of the specific car brand—stories the elements of which are aesthetically presented with overwhelming care and precision in the brand pavilions, and also in intellectually challenging ways, e.g. in the dioramas and omnimax presentations in the Skoda pavilion.

However, Wolfsburg, with Autostadt and ExperienceWorld, presents a Branding 2.0 model that aims at more than just developing a corporate brand. The aim is to develop a whole city and a region so that it may survive the changed economic and cultural

Figure 7. Audi pavilion behind Olafur Eliassons's Scent Tunnel
Source: Photo P. Allingham, April 2007.

conditions in a new Europe. The foundation of this is the PPP, Wolfsburg AG, with its network of business segments of which only a part of one, the ExperienceWorld, has been considered here.

Discussion and Concluding Remarks

Earlier, we have analysed how different methods of branding have been applied in experiential strategies to develop Dresden and Wolfsburg, two German cities. Before turning to the question of whether these examples present possible models of survival that might be adopted by other cities in a similar situation, the results of the analysis will be summarized.

At Dresden the branding model applied has been a corporate "Old School" branding model. The aim of its strategy has been to merge reciprocally the images of VW's assembly plant, Die Gläserne Manufaktur, and the luxury limousine produced there, the Phaeton, with Dresden's cultural Baroque image so that the purchase of the car has become the acquisition of a cultural image.

At Wolfsburg we found that the experiential vision had been structured by two major strategic designs: an architectural brain-and-neuronal-network-metaphor at AutoStadt and a topographic "hub-and-spokes" matrix for the experiential organization of the city and its surroundings. Both designs support the type of branding called Branding 2.0 above that stimulates and catalyses the creative ideas and movements of customers and guests, and plays down traditional commercial promotion. The novelty of this method is the apparent consideration for, and exploitation of, the individual guest's physical, intellectual and imaginative involvement on his or her own premises. Furthermore, the novelty is the step-by-step implementation of an experiential network strategy that organizes the activities of an entire city in a restructuring process with the point of departure in endogenous industrial legacy.

An evaluation of the branding methods applied in the development of the two cities must consider the fact that, although cars are produced in both cities and place-bound generated qualities enter into the experiential strategies, the two cities are very different. Therefore, there is reason to believe that what runs well in one case might be inefficient in other cases and that, consequently, it will be impossible to evaluate.

However, the analyses have pointed to a problem which is apparently shared by both of the applied branding methods. The problem occurs when aesthetic presentation secedes to such a degree that the connection to what has been represented becomes indistinct. This seems to have happened in the application of both branding methods, in which the application of aesthetics has been exaggerated in different ways. In the case of Die Gläserne Manufaktur and Dresden, the problem pointed to was aestheticization or simulation, and in the case of Wolfsburg/Autostadt it was sense-appealing aesthetics verging on "anaesthetics".

Therefore, returning to the advice of Pine and Gilmore of "wrapping experiences around existing goods and services", we must now conclude that wrapping alone is not enough.

In a recent publication, the authors have addressed the problem of representation in experience economy under the heading of authenticity. Their point of departure is the observation that "in a world increasingly filled with deliberately and sensationally staged experiences [. . .] consumers choose to buy or not buy based on how real they perceive an offering to be" (Gilmore & Pine, 2007, p. 1). Consequently, the authors recommend that businesses should base their strategies on authenticity.

In brief, Gilmore and Pine understand authenticity as socially constructed, i.e. what is authentic depends on what individuals perceive as authentic. Therefore, authenticity might be obtained through customizing appeal to the self-images of consumers (Gilmore & Pine, 2007, p. 12). However, authenticity also depends, on the one hand, on the fulfilment of two principles, viz. being true to yourself and being what you appear to be (Gilmore & Pine, 2007, p. 96) and, on the other hand, the type of economic offering. On the basis of five types of economic offerings: commodities, goods, services, experiences and transform-ation, the authors list five genres of perceived authenticity (Gilmore & Pine, 2007, p. 46). Experiences, they claim, render referential authenticity, i.e. "people tend to per-ceive as authentic that which refers to some other context, drawing inspiration from human history, and tapping into our shared memories and longings; not derivative or trivial" (Gilmore & Pine, 2007, p. 50). Obviously, this might, for instance, be some kind of memorable cultural legacy or heritage.

In view of this, the Dresden experiential model, the corporate branding model, at first seems well established in the "Manufaktur" concept referring to the Baroque epoch. However, the authenticity effect created through the reference to pre-industrial manufac-ture is contradicted by the apparent fact that Die Gläserne Manufaktur is a high-tech indus-trial production unit based on advanced automation and robotics. Furthermore, it offers its customers an educated cultural identity through the acquisition of a luxury car. However, the car brand is Volkswagen, a brand that generally, at least in Germany, is not associated with cultural education or luxury. If you add to this the Baroque equation between city and factory, as analysed above, it seems that credibility and authenticity is a major problem for both city and factory. Therefore, with Ruby's Las Vegas point above in mind, it seems that developing a corporate as well as an urban identity on nostalgic dreams of a selected glor-ious past probably has a limited sustainability.

My conclusion is that in the case of Dresden the corporate branding model seems a blind alley. This conclusion is further supported by recent criticism of the tendency of this model towards standardization, narcissistic self-absorption and a possible ensuing loss of reality (Christensen & Morsing, 2005, p. 136).

In Wolfsburg, the authenticity that might be reclaimed from the city's industrial and his-torical KdF-legacy has for obvious reasons not been included in the experiential strategy of either VAG or Wolfsburg AG. However, this evasion stirs the historical knowledge and memories of guests that may wonder how the legacy has actually been appropriated.

At first, the brandhub/wheel model applied in Wolfsburg seems to make up for the deficiencies of the corporate branding model. As a piece of neuronal or process architec-ture, Autostadt offers a catalysing frame of inspiration and experimenting, in which cus-tomers and visitors may become prosumers that invest and develop their individual ideas. These ideas and projects can be further extended and developed in the network of experi-ential nodes of Wolfsburg outside Autostadt, provided that the possibilities of doing so have been communicated clearly and that a sufficient number of corridors between the experiential nodes are available.

However, a number of problems have been detected in the design of the Autostadt brandhub. The strong foregrounding of sense-appealing aesthetics expressed in layout, architecture, design and art can be understood as an intervention that diverts, controls and channels the projects of visitors and guests to the extent that it opposes and weakens the creative implements of the Branding 2.0 method, making it resemble Brand-ing 1.0. In addition, the co-branding element considered as one of the characteristics of brandhubs is in the case of Autostadt not "genuine", as all the brands are owned and managed by one operator, the VAG (cf. Höger, 2004). Anyway, the Branding 2.0 model seems to be more in accordance with a contemporary constructionist receiver-oriented experiential strategy than Branding 1.0, as it renders authenticity through customizing appeal to the self-images of the visitors. Therefore, it represents a better strategic model of survival that might be adopted by other cities with the following provisions.

First, the presence of a historic legacy, a weighty economic background and place-bound cultural resources seems vital. Secondly, local commercial and/or public operators with identical aims must be present and willing to cooperate. A governance structure, e.g. a PPP with a common vision, seems prevalent to promote local and infrastructural development. Such partnerships are behind emerging hybrid urban neo-corporativist brand- and entertainment-scape types such as Sony Center or Daimler City in Berlin. Sony Center in Berlin is also characterized by co-branding and cooperation between public and private organizations like cinemas, museums, VW, Sony, Lego and many more.

In general, it seems that authenticity derived from place-bound legacy and cultural heritage plays a major role in experiential strategies for developing mid-sized and small cities and regions. In an article on the redevelopment of the Ruhr District, Professor of Strategic Management at Saxion Universities for Applied Sciences in Enschede in Holland, Gert-Jan Hospers (2004, p. 155), stresses that place-bound history and endogenous industrial legacy have shown the way to a new economic future for many cities in the Ruhr District. He refers to successful examples of city branding projects like "E-city Dortmund" and "Solar City Gelsenkirchen". These cities are, according to Hospers, examples where balances have been found between experiential means and methods and regional industrial traditions.

Such a point seems to be in accordance with Gilmore and Pine's Here-and-Now Space model for an authentic strategy process. The first step of their eight-step model is "Study your heritage" (Gilmore & Pine, 2007, p. 189).

Note

1. On orthography: I translate all Danish and German quotations, although proper names and some slogans will appear in German. The special spelling of proper names used by Volkswagen AG (henceforth VAG) and Wolfsburg AG like "AutoStadt" has been retained both in German and English instances.

References

Ahrens, K. (2001) Denkmal für Dresden, *Managermagazin*, 4, 1 April, pp. 202–209.

Buhl, C. (2005) *Det lærende brand* (København: Børsens Forlag).

Christensen, L. T. (1994) *Markedskommunikation som organiseringsmåde* (København: Akademisk Forlag).

Christensen, L. T. & Cheney, G. (2000) Self-absorption and self-seduction in the corporate identity game, in: M. Schultz, M. J. Hatch & M. H. Larsen (Eds) *The Expressive Organization*, pp. 246–270 (New York: Oxford University Press).

Christensen, L. T. & Morsing, M. (2005) *Bagom Corporate Communication* (København: Forlaget Samfundslitteratur).

Eckhardt, E. (Ed.) (2002) *Merian—AutoStadt in Wolfsburg* (Hamburg: Jahreszeiten-Verlag).

Eco, U. (1979) *The Role of the Reader* (Bloomington, IN: Indiana University Press).

Eco, U. (1990) *The Limits of Interpretation* (Bloomington, IN: Indiana University Press).

Foucault, M. (1980) *Power/Knowledge* (Brighton: Harvester Wheatsheaf).

Gilmore, J. H. & Pine, B. J., II. (2007) *Authenticity: What Consumers Really Want* (Boston, MA: Harvard Business School Press).

Gingco.net Werbeagentur (2005) *Die Gläserne Manufaktur—Die Vision wird Realität* (Dresden: Automobilmanufaktur Dresden GMBH).

Grant, J. (2006) *The Brand Innovation Manifesto: How to Build Brands, Redefine Markets and Defy Conventions* (Chichester: John Wiley).

Hamilton, G. (Ed.) (2007) *Guidebook on Promoting Good Governance in Public-Private Partnerships* (New York and Geneva: United Nations).

Hatch, M. J. & Schultz, M. (2004) The dynamics of organizational identity, in: M. J. Hatch & M. Schultz (Eds) *Organizational Identity: A Reader*, pp. 377–403 (New York: Oxford University Press).

Henn Architekten Ingenieure (Ed.) (2000) Corporate Architecture, (Berlin: Aedes).

Höger, K. (2004) Brandhubs—catalysts for responsive urban design, in: K. Christiaanse (Ed.) *Entwurf und Strategie im urbanen Raum—Die Programmlose Stadt*, pp. 57–75 (Zürich: Professur für Architektur und Städtebau).

Hospers, G.-J. (2004) Restructuring Europe's Rustbelt, *Intereconomics: Review of European Economic Policy*, 39(3), pp. 147–156.

Iser, W. (1974) *The Implied Reader: Patterns of Communication in Prose Fiction from Bunyan to Beckett* (Baltimore, MD: The Johns Hopkins University Press).

Jakobson, R. (1971) *Selected Writings*, Vol. II. (The Hague: Mouton).

Jantzen, C. & Vetner, M. (2006) Oplevelse—Et videnskabeligt glossar, in: C. Jantzen & J. F. Jensen (Eds) *Oplevelser: Koblinger og transformationer*, pp. 239–260 (Aalborg: Aalborg Universitetsforlag).

Kunde, J. (1997) *Corporate Religion* (København: Børsen Bogklub).

Lincoln, Y. S. & Guba, E. G. (1985) *Naturalistic Inquiry* (Beverly Hills, CA: Sage).

Mollerup, P. (1997) *Marks of Excellence: The History and Taxonomy of Trademarks* (London: Phaidon).

Olins, W. (1989) *Corporate Identity: Making Business Strategy Visible through Design* (London: Thames and Hudson).

Peirce, C. S. (1931–1958) *Collected Papers*, Vols. 1–6. (Cambridge, MA: Harvard University Press).

Pine, B. J., II & Gilmore, J. H. (1999) *The Experience Economy: Work Is Theatre and Every Business a Stage* (Boston, MA: Harvard Business School Press).

Pohl, J. (2005) Urban Governance à la Wolfsburg, in: A. Kübler *et al.* (Eds) *Informationen zur Raumentwiklung*, Heft 9/10, pp. 637–647 (Bonn: BBR).

Rauterberg, H. (2001) Ganz grosse Fragen am Fliessband, *Zeit Online*, 51(13.12), pp. 42–44, Available at http://www.zeit.de/2001/51/Ganz_grosse_Fragen_am_Fliessband (accessed 20 March 2009).

Research Group MAERKK (Market Communication and Aesthetic, Reception, Culture and Cognition) Aalborg University (2009). Available at http://www.maerkk.aau.dk (accessed 21 March 2009)

Riewoldt, O. (2002) *Brandscaping—Worlds of Experience in Retail Design* [Erlebnisdesign für Einkaufswelten], (Basel: Birkhäuser—Publishers for ArchitectureVerlag für Architektur).

Ruby, A. (2000a) Gläserne Manufaktur Dresden. Das Unternehmen in öffentlichen Raum, in: Henn Architekten Ingeneure (Ed.) *Corporate Architecture* (Berlin: Aedes).

Ruby, A. (2000b) Las Vegas an der Elbe, *Die Zeit*, Ausgabe 46, p. 59. Available at http://www.zeit.de/2000/46/Las_Vegas_an_der_Elbe (accessed 21 March 2000).

Stigel, J. & Frimann, S. (2006) City branding—all smoke, no fire? *Nordicom Review*, 27(2), pp. 243–266.

Wolfsburg AG (Hrsq.) (2007) *A Vision with definite prospects* (Braunschweig: Maul-Druck GmbH).

Wolfsburg Marketing GmbH (Hrsg.) (2006) *Schauen, Entdecken, Erleben Wolfsburg* (Wolfsburg: Wolfsburg Marketing GmbH).

The Experience Economy and the Transformation of Urban Governance and Planning

HANS PETER THERKILDSEN, CARSTEN JAHN HANSEN &
ANNE LORENTZEN

ABSTRACT *This article discusses the relationship between experience-oriented development and urban governance and planning, based on a case study of the city of Frederikshavn (DK). Triggered in 1999 by a sudden local economic crisis, Frederikshavn entered a process that reinvented its "mental frame" and transformed not only its urban development, but also its identity, image and governance towards an experience economic and entrepreneurial profile. We investigate what influenced urban strategy-making and planning in Frederikshavn and allowed the city to move towards an experience economy. Municipal investments, internal reorganization and public–private cooperation played significant roles. Traditional spatial (land use) planning and regulation were replaced with transformative urban growth strategies and more risk-taking experimental approaches. The municipality became a project partner that favours "actions because they create new opportunities". Experience-oriented projects thrived in this entrepreneurial environment. However, recent political tensions between growth and welfare agendas indicate that Frederikshavn thereby exemplifies a test to the reaches or limits to government-supported neoliberal approaches in urban development and governance—and thereby also to the role of the local state.*

Introduction

Many western-world cities are reorienting themselves in order to strengthen their development conditions, often in response to local industrial decline as well as intercity and global competitiveness agendas. Recently, the "experience economy" has emerged as a post-industrial development perspective that may provide a new or different kind of agenda and dynamics to processes of change in urban areas (Schulze, 1992; Clark, 2004; Romein, 2005; Lorentzen, 2009). Alongside this, urban governance and planning activities increasingly seem to be moving away from simplified physical views of cities to a much more integrated and strategic approach, in which attention is focused on the complex inter-play of economic, socio-cultural, environmental and political/administrative dynamics as these evolve across and within an urban area (Healey, 2007, p. 3).

It is the main purpose of this article to discuss how the experience economy may relate to urban governance and planning. How do activities and values embedded in experience-oriented urban development correlate with changes in urban governance and planning settings, modes and roles? Will increased attention to experience-oriented development reflect in new more integrated and strategic approaches in urban policymaking and planning, and vice versa? Will experience-oriented activities require or lead to new constellations and interaction patterns of actors and interests? The article searches for lessons to be learned that may prove helpful in conceptualizing these relationships.

In doing so, we explore the example of the Danish, the city Frederikshavn (24,000 inhabitants) in which significant changes have occurred during the last decade, both in development conditions and in urban governance and planning. Frederikshavn is in transition from being dominated by traditional industries, primarily related to the sea port, towards economic development and public and private initiatives that increasingly focuses on experience-oriented aspects. Moreover, the Municipality of Frederikshavn has developed its own variety of tailored and more entrepreneurial approaches to urban planning where holistic plans are replaced with project approaches and more strategic uses of participation. It is exactly this possible interplay between new developments in urban governance and the transition to a more experience-based economy which com-prises the focal point of the article.

However, before exploring the Frederikshavn case the following two sections will attempt to characterize how the experience economic development perspective and recent changes in urban governance and planning may be perceived.

Experience Economy and Urban Development

The "experience economy" has emerged as a term to describe a new economic era based in the added value of experiences, and it is powered by the increasing affluence, competition, technological development and personal engagement of the customer (Pine & Gilmore, 1999, pp. 1–6). Citizens of more affluent societies seem to demand, and spend an increas-ing amount of resources on, music, sport or art events, museums, theatres, heritage sites, festivals, etc. This may not only be seen as an expression of Clark's (2004) statement that the "consumer drives the modern economy" (p. 20), it also indicates that experiences are becoming increasingly important to the formation of social and, in particular, individual identities (Romein, 2005; Hall, 1998). Citizens demand that interesting, memorable and identity-creating experiences and events become a more integral part of everyday life. As such, the growth in the production and consumption of experiences may also be seen as an example of the emergence of a "post-service economy", as predicted by Toffler (1970), or the experience society (Die Erlebnisgesellschaft) as termed by Schulze (1992).

Cities serve as centres for the consumption of events and other experiences. A new vocabulary is emerging on how experience-oriented activities may relate to urban devel-opment, e.g. Experiencescapes (O'Dell & Billing, 2005), Fun city (Marling & Zerlang, 2007), The city as an entertainment machine (Clark, 2004), The event city (Metz, 2007), Fantasy city (Hannigan, 1998). In any case, it indicates that experiences may change the function and physical expression of cities. The infrastructure of material production (factories and warehouses) that visually and spatially dominated the landscape of the industrial city has been replaced, not only with the glittering office complexes of the post-industrial service economy, but also, and to an increasing extent, with entertainment sites, districts and events (Romein, 2005, p. 8).

Lorentzen (2009) suggests that places in themselves can be objects of consumption. Buildings, streets, squares, parks and other public spaces in cities can have certain well-designed or functional characteristics that make them attractive for people to visit, use or enjoy. Many experiences are "place-bound" (Lorentzen, 2009; Smidt-Jensen *et al.*, 2009), and consequently, the increasing role of experiences can often be identified as specific changes in the urban physical landscape. In addition, experiences seem to constitute an increasing part of urban social and economic life. Their quality therefore becomes a factor to urban development. Some scholars even argue that in the post-industrial city, the relationship between urban economy and urban design has reversed: while the quality of urban environment was an outcome of economic growth of cities, it has now become a prerequisite for economic development in the most direct manner because it lures consumers (Romein, 2005, p. 10).

The experience offer of a city can be seen as a part of its "quality of place" (Florida, 2002). Florida's basic argument is that the quality of place decides where highly educated people locate. This may be dependent on: the physical and functional combination of buildings and the natural environment, the diversity of people (including the presence of entrepreneurial minds with the capacity and creativity to be innovative in terms of experiences and be able to stage them) and, finally, specific activities and events that embed or signal a vibrant city life with cultural, exciting and creative content (Florida, 2002, pp. 231–234). In line with this, Clark (2004, p. 106) suggests that the provision of urban amenities can be seen as a driver of growth and a factor to the location of human capital. According to Clark, urban amenities include natural amenities (e.g. climate, water), constructed amenities (e.g. museums, coffee bars) as well as socio-economic composition and diversity and values and attitudes of residents.

In this article, we are particularly interested in the quality of place that relates to the experience economy and the experience society. As the demand for experiences grows, the importance of recreational and leisure-oriented facilities grow in relation to urban development and planning—compared with the demand for the traditional fields of housing, educational and health and social-care facilities. Experience offerings and amenities become a means for urban public officials and businesses to enhance the quality of their locations in the eyes of present and future residents, tourists, conventioneers and shoppers (Clark, 2004, p. 1). Such changes in development imply that there may be new challenges emerging to the governance and planning of cities.

Experience Development and Urban Governance and Planning?

The production of leisure-oriented places and activities in urban areas is becoming increasingly market-driven (Romein, 2005). In the 1950–1970s, such activities and places were most often supplied by the public sector as part of welfare programmes (e.g. resulting in museums, parks, sports facilities, music halls, etc.). From the 1980s and onwards, however, welfare states tend to have "slimmed" and increasingly left the production of such activities to the market. In addition, market actors tend to have discovered the potential in cultural symbols to become a lever for expanding consumerism (Romein, 2005, p. 12).

This could imply a reduced importance of experience-oriented aspects to urban public policy-making and planning. This is however, not the case. On the contrary, strategies aiming at building experience-oriented places and activities seem to be abundant in present-day urban public policies, strategies and plans.

The conception of cities is often understood as an economic, political and cultural unit that must implement activities in an entrepreneurial spirit in order to strengthen its competitiveness. This transition of the self-image of cities is closely connected to the transition of governance, particularly in relation to the new types of public–private partnerships and networks (Jessop, 2000, p. 93). In addition, Brenner (1999, 2004) has used the term "glocalization" to indicate that cities and regions (rather than nations) are principal actors in territorial competition and the ongoing restructuring related to a globalizing economy—and in order to reconfigure themselves, cities tend to develop new governance modes that recombine and aggregate public and private resources. Former bureaucratic welfarist approaches are being replaced by neoliberal political and economic growth agendas and entrepreneurial approaches of the local state. Entrepreneurial urban government and planning has turned from regulating and redistributing urban growth to encouraging it "by any and every possible means", in particular, by luring prospective people and firms (Romein, 2005, p. 13). In general, entrepreneurial urban governance has correlated with an increased attention to make more efficient the (local) welfare state expenditure, while, at the same time, the responsibility for urban revitalization and growth is expanded well into the private and civil sphere through partnerships and participation initiatives.

These changes may also be seen as part of a fundamental shift in the processes through which local-level urban change is planned, negotiated and implemented (Elwood, 2004, p. 755). Local-level policymaking and planning activities seem to have changed in ways that have increasingly been characterized using terms of dialogue, cooperation and networks (Hansen, 2006). In general, it is claimed that hierarchically organized institutions increasingly find it difficult to handle contemporary and often rapid social, technological and economic changes through schematic top-down regulatory approaches. In response to an apparently limited range of predefined approaches in governing, it seems that new, more informal and often *ad hoc*-oriented practices for collective action have been gaining ground. Such practices have been termed dynamic or fluid networks, in which there is a focus on collaboration and the coordination and pooling of public and private resources, as well as a focus on the establishment of more situation-specific "rules-of-the-game" and problem- and project-oriented approaches (Hajer & Wagenaar, 2003; Bogason *et al.*, 2004; Kooiman, 1993; Dryzek, 2000).

Hence, the idea of entrepreneurial urban governance is accompanied by discourses concerned with partnerships and often new ways of participation in urban planning and revitalization processes (see also Elwood, 2004). In addition, Healey (2007) emphasizes that

> the "places" of cities and urban areas cannot be understood as integrated unities with a singular driving dynamic, contained within clearly defined spatial boundaries. They are instead complex constructions created by the interaction of actors in multiple networks who invest in material projects and who give meaning to the quality of places. (p. 2)

Therefore, traditional approaches to urban governance and spatial planning are left behind. "The planning project" becomes "a governance project" (Healey, 2007, p. 3), in which local actors as well as institutions assume new settings and roles.

The discussion of the production of experience-oriented places and activities in urban areas is closely related to emerging understandings of how strategies and urban spaces are produced and reproduced by local states and the shifting array of actors and institutions that are engaged in governance at this level. A focal question becomes: what constellations and interaction patterns of actors and interests may match strategy-making for experience-oriented places and activities? In a broader attempt at debating new urban governance developments, Pløger (2004, p. 178) tentatively conceptualizes a strategic urban govern-ance that puts complexity and flexibility at centre-stage and attempts to make space for enveloping urban culture, identity and meaning. According to Pløger, and based in Aschér's (2002) views on "meta-urbanism", such an approach:

– involves a new urban management that defines and works with different projects in the city and attempts to create a local coherence around those projects. This approach aims to take into consideration unforeseen events along the way—it lives with risks in its practice without prefixed procedures for reducing risks.
– still celebrates efficiency and prioritisation. However, it is public and private actors and interests in unison who must find ways of realisation that are the most efficient for the community and for the parties involved. Urban management must not simplify complex realities, but work with them instead. Efficiency and durability is better achieved through variation, flexibility and reactivity. This urbanism must work with varying and changing needs and with a continuous reconfiguration of interests (Pløger, 2004, p. 178, our translation).

In this approach, urban governance, management and planning are seen as based in projects, and hence in contextuality, actuality and locality. It is sensitive to the field of actors and interests, and it does not reduce public enquiry and debate to predefined and specific phases. The political and centralized governing based in bureaucracy and insti-tutional power is replaced with an open, dialogue-oriented, self-reflective, transformative, micro-political and locality-oriented mode of management and governance (Pløger, 2004, p. 179, our translation). As such, this seems to exemplify a neoliberal approach that embeds and combines discourses and mechanisms of urban entrepreneurialism and of partnership and participation.

The question is then, how do new forms of governance emerge locally? And further, in which particular ways do they relate to specific challenges of urban development and change? For example, state structure has a tremendous impact on how and to what effect the engagement of local citizens in partnership and participation plays out (Elwood, 2004, p. 758). However, as Pløger indicates, a range of more context-dependent and situation-specific local conditions will also be likely to have an effect. For instance, the local transition from a crisis-stricken industrial production to an alternative business structure can be harsh, but it also constitutes a moment of opportunity. Healey (2007, p. 195) argues that such moments are suitable for engaging key stakeholders in strategy-making because the need for action is obvious. Furthermore, moments of oppor-tunity facilitate transformative potential in strategies while it is evident that change is required (Healey, 2007, p. 198).

In searching for a useful vocabulary to further characterize new ways of urban governance, Healey (2007, p. 30) points towards the possibility of seeing practices of urban governance in terms of more informal, less disciplined, instrumental and nonlinear versions of strategic thinking. Healey (2007, p. 30) suggests that "a strategy is better understood as a discursive frame, which maintains 'in attention' critical understandings about relationships, qualities, values and priorities". In this sense, a strategy needs to be flexible and strong in order to have the capacity to travel and adapt into a diverse range of complex institutional and discursive contexts without losing core values. The strategic frame is a selective social construction based on collective sense-making, and if success-ful, it motivates people through seductive and persuasive properties. A strategy that accumulates substantial persuasive power becomes a part of the structuring dynamics within which subsequent actions are embedded (Healey, 2007, pp. 184–186). Moreover, a strategic frame is shaped by the flow of action and it is not bound to be formulated before action is taken. In addition, a strategic frame should rely on and leave room for serendipity as much as a deliberate process of creation (Healey, 2007, p. 183).

This implies that strategic frames are not only the product of the state, but rather of a broader range of actors. Steering capacity emerges among a mixture of public and private actors, and it "is being carried through using persuasion, seduction and inducements" (Healey, 2007, p. 182), rather than by regulatory means. This demands great attention to how the strategy is imagined, presented and intertwined with daily political work and development efforts. Strategies generating transformative force "shape practices rather than specific decisions, through providing a different way of making sense" (Healey, 2007, p. 195).

In this article we apply the notion of strategic framing as a way to characterize the emer-gence of new institutions and discourses in governance and planning. It is a notion that implies a key role to be played by actors and stakeholders and the networks that may emerge between them. The dynamics and transformative power of such networks is con-sidered to be dependent on their composition, processes and situational context. Moreover, the influence of actors and networks is not necessarily related to traditional decision making power, although old power structures may continue within the new strategic frame. It is not crucial to actually possess the power to make decisions. What is more important is to make arrangements and "organise coalitions that cut through the diverse governmental agencies to create visions with which people can work" (Hajer, 2005, p. 641). Such arrangements and coalitions—or what we choose to broadly term "governance networks", because we assume them to contain steering or decision making capacity—may be tailored for specific purposes, such as the development of shared visions or common understanding, the building of legitimacy or loyalty, as well as the building of a better basis for effective and joint implementation.

However, governance networks typically function in the absence of clearly defined constitutional rules (Hajer & Versteeg, 2005, p. 340), which may cause challenges to democratic decision making and legitimacy. On the one hand it "may create new spaces for democratic practice and afford citizens and community organizations a wider array of strategies through which to influence urban decision making processes and legitimize community contributions in these processes" Elwood (2004, p. 756). Or it may be seen as "a disempowering strategy for disciplining citizen participation into certain acceptable forums, limiting citizen voice to particular arenas or removing a basis for resistance to state agendas" (Elwood, 2004, p. 756). This brings into attention: what kind of actors are involved in governance networks? what are the relations between the participants? how do internal and external coordination take place? and how legitimate are the strategies evolving from the different practices of the networks?

Moreover, Metz (2007, p. 30) argues that positioning a city as a centre for events can cause unpleasant situations for the local government, such as having to choose between entrepreneurs and citizens. Experience-based growth policies may conflict with redistributive welfare policies. On the other hand, as there is a demand for experiences in the public, the interests of entrepreneurs and citizens are not necessarily conflicting. Also, new experience-based amenities may not threaten local government welfare services if they are either part of private enterprise or if they, as public investments, increase the overall urban economy (and local tax revenues).

Given the mixed possible outcomes of the described developments, it seems critical to investigate and discuss urban governance and planning related to the development of experience-oriented places and activities. This will be carried out through a case study of Frederikshavn. In the case study we specifically ask: what has influenced developments in urban strategy-making and planning in Frederikshavn and allowed the city to move towards an experience economy? In general, we focus on the emergence and handling of (new) ideas, interests, actors, networks and activities, and the conditions and opportunities shaping these. Furthermore, we search for the flexible, informal and nonlinear processes and practices in the urban governance of Frederikshavn and critically discuss legitimacy and the possible tensions between the welfare and growth policies of the city.

Our methodological approach is post-positivist interpretative (Hajer & Wagenaar, 2003; Flyvbjerg, 1998). This essentially means that the concepts presented above were not chosen prior to the case study. Rather, this conceptualization emerged from an iterative process; informed by a dialogue between pilot-studying Frederikshavn and studying the literature. In the case analysis this methodological approach has led us to focus specifically on: the emergence and interaction of urban development values, visions, aims, strategies and activities, as well as the associated settings and ways of collective and strategic problem-solving and public–private cooperation. Furthermore, we have investigated: changes in urban economic activities, in particular, place-bound experience-oriented activities, changes in the spatial landscape, changes in urban policies, strategies and plans and changes in constellations and practices of actors and networks. The empirical base consists of public planning documents, reports, newspaper articles and 12 qualitative interviews—marked (x) below— and conversations with political and administrative stakeholders, representatives from the local business community and citizens (Therkildsen, 2007).

The Frederikshavn Case

> Some call it a change process. Mostly, we call it a transformation process. Transformation is when something takes on a new appearance, a new expression, a new identity. To a high degree, the municipality has been the driver of the transformation, but it has happened in very close cooperation and with a strong will towards the potential for growth in the local community.[1]

Frederikshavn is located by the sea in the northern part of Denmark, with a naval base and ferry routes to Norway and Sweden. Recently, the city has undergone considerable changes that may seem rapid and surprising, as the traditional industrial and maritime town transformed itself, within a brief span of time, into what may be termed an "experience city". We consider Frederikshavn to be a critical case of the transformation of urban governance and planning in relation to the experience economy.

The Crisis and the Development That Followed

In 1999, the largest private sector employer in Frederikshavn, the "Danyard" shipyard, closed down. The same year, duty-free shopping on ferries to Norway and Sweden was brought to an end due to a decision by the European Commission. Together, this resulted in the loss of more than 5000 local jobs, primarily in retail and the maritime sector. It was a devastating blow to the local labour market and to the identity of the city, having seen itself as a "shipyard city". In the years that followed, a wide range of specific projects was implemented or initiated in Frederikshavn. These projects have helped to restore the local employment situation and significantly influenced the development and business profile of Frederikshavn.

Following the collapse of Danyard, from 1999 and onwards the municipal council initiated public investments in urban public spaces and several cultural, sports and experience-oriented facilities. A first step was to improve the worn down industrial physical image of the city centre by beautifying squares, pedestrian areas and streets. Later, this included the development of different paving and lighting. Local culture and leisure were favoured further by the municipality through the building of new facilities, such as the extension of the ice stadium in 2003, a new music house in 2004 and the combined sports and cultural arena "Arena Nord" in 2005 (Arena Nord). Furthermore, in 2004, a rather unconventional municipal project became a reality—the establishment of a palm beach in proximity to the city centre, with imported palm trees and added white sand. The beach is maintained by the municipality, e.g. the trees are moved indoor each winter.

Apart from investment in buildings and facilities, a range of new experience-oriented activities has emerged (Lorentzen, 2007, 2008), mainly based on private sector sponsorships. In 1998, the Festival of Tordenskiold (celebrating eighteenth century sea port life and a sea battle) was initiated. It was initially a small local event that involved local citizens. However, it has now become a larger annual event and attraction with more than 25,000 visitors and 1000 local participants and activists (Lorentzen, 2008, pp. 10–11). In 2004, the first light festival was held, and in 2006 it developed into an international festival, Light Visions, with light artists from more than 15 countries. It is now a continual event. This has spurred the Light Visions Innovation Centre, established in 2006 by a wide range of actors, mainly business. Several other significant events have been arranged, e.g. concerts, conventions, sport championships and conferences—culminating with visits by Bill Clinton in 2006 and Al Gore in 2007.

These projects have contributed to changing the development and business profile of Frederikshavn from industrial production (mainly) towards a significant increase in the production of experiences. Meanwhile, the businesses related to the harbour have become more oriented towards maritime services and knowledge than earlier. For instance, "Frederikshavn Maritime Erhvervspark" (Business Park) has been steadily expanding to contain more than 60 smaller and mid-sized companies, primarily within maritime service and technology.[2]

Overall, through 2000–2006, approximately 2.2 billion DKR (approx. 235 million £, per 22nd September 2008) has been invested, both public and private, in buildings and facilities for business, culture and experiences.[1] The municipal investments in urban gentrification and leisure and cultural facilities have been considerable compared with the size of Frederikshavn and to earlier local public investments. This is primarily due to hard prioritization in the municipal budget, however, in some cases also due to loans and EU co-funding.[1,3]

The renewed development and business profile of Frederikshavn has also been mirrored quite strongly in the media. Around 2000, Frederikshavn was mostly characterized in relation to negative issues such as unemployment, shipyard crisis and violence. However, through the 2000s this changed dramatically. The city was mentioned more, and increasingly for positive aspects such as creativity, attractiveness, courage, enthusiasm, renewal, etc. In particular, the word "experience(s)" entered the headlines to a significant extent (Lorentzen, 2007). Alongside this, the self-image among the citizens of Frederikshavn also seems to have changed dramatically in a similar positive direction, from hope and wary optimism to pride and confidence (Lorentzen, 2007).[1-5]

How and why did all this come about? This will be discussed in the following sections.

Changes in Mentality and Identity

The problem was not the closure of a ship yard. The problem was that the kind of production represented by the ship yard did not have a future. ....then, what could happen in our city, so that mentally we would be able to have other forms of production? That was what we started thinking about. ...[the Danyard closure] was the best that ever happened to us. Absolutely. Without a doubt. Because, it has forced us to think differently.[3]

The Danyard closure was not entirely unexpected. The shipyard crisis had emerged during the 1980s, both in Frederikshavn and as a tendency in many western cities. However, in Frederikshavn the issue of a possible closure became a taboo; it was not to be mentioned, and "not even to be thought about, because it might make it real".[3] The fear of losing jobs produced a political and developmental gridlock that locked the business and urban development profile of Frederikshavn and made it practically impossible to discuss an alternative future without shipyards.[3,4] This also meant that, when it happened in 1999, the closure struck hard because "there was no contingency plan".[4] An immediate sense of failure and urgency set in among local politicians, public servants and business life as well as the public. It was characterized as "a loss of identity"[4] and there was a widespread recognition that earlier approaches had to be abandoned.[3]

This mood soon developed into a "refusal to blackout",[1,4] an atmosphere of "having the back against the wall together",[1] and that common action had to be taken. The head of the municipal Technical Administration elaborates: "The mentality started to change ... The reality that hit us was that if we didn't do something now, and if we did it the traditional way, then we would be dead. ....Then we realised that when we did something, it might not be completely right, but it opened doors that we did not know were there. So, our action created the occasion for something completely different. ...by doing something we opened up something that we didn't know was there, and that was where we could make a choice".[3]

Among key politicians, public servants and business representatives an entrepreneurial spirit and more risk-oriented mindset started to flourish that encouraged and favoured "the ability to image things differently than they are; the ability to see things in a new perspective".[1] It was partly provoked by the Danyard closure, and partly by a common recognition of "the spirit of the times"[3] and of being "part of a globalized world".[1,2] In addition, it was due to local values of being "somewhat fisher-like, devil-may-care, a bit disorganised and Klondike-like, which has perhaps formed this place for generations".[4] It allowed new ideas and visions, as well as a range of specific actions, to emerge.

> We wanted to become the maritime centre of Denmark. It took some strength because we had just failed utterly. ... We said to each other: people get richer. and more leisure time while becoming more stressed. They want experiences, and they want contents in their leisure time. To create that, in principle one had to build it as an industry. ...We also said: we are used to tourists and guests; shopping guests from Norway and Sweden. We know how to service people. ...We can see that experience economy or events; that there is a lot of money in that.[4]

As such, the ideas and visions that grew had references to the maritime and one-day-visit (tax-free) tourism past of Frederikshavn. However, while the intention evidently was to "respect our history"[1] it was equally clear that this process was supposed to lead to a rein-vented and "more modern identity and development".[3,4] It meant the replacement of identity-shaping keywords such as "industry" and "tourism" with "service" and "experience". It also included a growing local sense of ability to be able create a role in a globalized devel-opment, or as it has been formulated by locals: "Denmark is too small for Frederikshavn".[3]

These changes initiated the formation of a renewed local "mental frame"[1,3] that soon influenced the emergence of the wide range of activities and projects mentioned above. However, it also became clear that specific action (projects) carried the potential to change not only (back on) mentality and the city's identity, but also strategy-making and planning as an activity. Before discussing that further, we first turn to clarify the con-tents of the new strategies that emerged and how the processes of change were organized and carried out.

New Strategies, Plans, Organization, Processes and Roles

The crisis situation, and the renewed mentality that grew from it, was followed by signifi-cant changes in local strategies and plans. First of all, earlier plans were quickly considered to be of no use, as they offered no visions for a future without shipyards. In example, the 1989 municipal plan made priority to "broaden the business community on the basis of the existing industries" (Frederikshavn kommune, 1989, p. 30). It was recognized that the business structure was narrow and sensitive to the state of the market, but an alternative business structure was not part of the plan.

There was no rush, in 1999, to fill in the planning void with new plans. The main attention was given to the development of specific projects, such as the urban renewal initiatives. However, together with local business, the Ministry of Business, the county and several more, the municipality devised a business-oriented strategy in 2000 (The Frederikshavn Area Towards 2005). The strategy was not compulsory, but emerged as a common initiative between the involved parties. Frederikshavn was to remain a strong player in the maritime field and tourism, although with a renewed focus. Furthermore, the intention was to increase the educational level and change the prevalent attitude in Frederikshavn from an industrial employee-culture to an entrepreneurial culture. In prac-tice, it became a biped strategy that focused on the maritime service sector and the cultural and leisure-oriented sector.[4]

In 2001, a compulsory municipal plan stated that the aim of local government was to create "fertile entrepreneurial environments and motivate the individual's inclination and ability of private enterprise" (Frederikshavn kommune, 2001, pp. 1–3). Moreover, the plan questioned the efficiency of traditional planning tools in enabling development. "We know that planning is not enough. Ideas and projects shall be born and realized among the citizens. Planning can provide a helping hand, but initiatives must be rooted in the daily life of the area to result in fruitful results" (Frederikshavn kommune, 2001, pp. 13–14). Traditional and regulatory land use planning was reduced to a minimum that barely fulfilled the requirements of the Danish Planning Act.[3,4] The local downscaling of land use planning and increased attention to economic development strategies were not only influenced by the local crisis. It was also seen as part of "the spirit of the times"[3] in Danish spatial planning. During the 1990s, urban economic growth strategies had emerged in many Danish municipalities as a deliberate supplement to compulsory municipal land use planning. It reflected a change in municipal political agenda from welfare distribution towards a more business and economic development-oriented approach. (Sehested, 2003) Moreover, from 2000 the Planning Act was altered so that municipalities were obliged to produce a "plan strategy" (containing overall visions and indications of actions) in addition to the municipal plan.

In Frederikshavn, the local crisis stretched these tendencies to their legal limit. The municipal council and key public servants became mostly concerned with building investments and tailoring specific projects together with local business life. Traditional regulatory land-use-oriented spatial planning was "pushed downwards" towards the project level. To compensate for cursory spatial planning, the municipal politicians and administration introduced internal ongoing scenario discussions[3,5] that intended to ensure a certain degree of preparation and readiness when specific projects emerged. In 2005, this resulted in "something that turned out to become a strategy: From Ship Yard City to Host City. This soon became an expression of the transformation process that was emerging".[1] Furthermore, in 2007 a Project and Communications Division was established in the municipal administration with a cross-disciplinary, coordinating role in order to keep track of the wide array of projects. In addition, the Technical Administration (planning authority) has been reorganized "so that sections are precisely a bit too small to solve a task on their own. ... it means that at least two sections look at a task, from each their point of view. It only becomes more qualified from that. It may take a week longer, but it is better".[3]

The changes in internal municipal organization have been accompanied by the development of a range of new local networks, organizations and participatory processes between the municipality, business, actors within culture and education and citizens. Business life has been favoured in particular, for instance through the establishment of a municipal Committee for Business and Tourism, that includes not only key municipal council politicians, but also representatives from business, commerce, a workers union and an education institution. Usually, Danish municipal committees only contain municipal council members, however, the municipalities are allowed to also set up supplementary and more mixed committees that contribute to the formation of municipal policies. It is the task of the Business and Tourism Committee in Frederikshavn to "create conditions for innovation", "ensure cooperation and interdisciplinarity" and "contribute to the formation of policies and strategies" concerning business and tourism development. The joint effort of the business community and the municipality has been further developed through several more specific projects and networks, e.g. a partnership with local light firms (in order to equip the city with cutting-edge lighting solutions), Frederikshavn Event (that attracts world known names such as Clinton to the city) and a close cooperation with a communication company "Tankegang" that uses Frederikshavn as its "laboratory" for developing new communicative strategies.[5]

Apart from the compulsory and traditional hearings related to municipal plans, the broader involvement of stakeholders and the citizens has become more project-focused and oriented towards specific groups. Before initiating larger projects, a meeting is usually organized where key stakeholders are invited to participate. It is not necessarily the municipal administration that invites and heads the meetings. If the project has private initiators they will be in charge of the sessions, which sometimes create different settings and alternative modes of deliberation. Such meetings are not always open to everyone, and the purpose is most often to engage stakeholders who can contribute to the project. At the public library, the municipality has organized a room with models and descriptions of current urban development activities and future possible projects. Here, all citizens are encouraged to express their opinion to the local municipal politicians and administration, e.g. on "post-its". Another example is the UNGwelt (youth world) project, which is an umbrella for various kinds of projects, targeting the age group between 16 and 25. Here, workshops and events are used to bring together young people, business, education and the municipality in order to increase the attractiveness of the city to young people. In sum, the extra-compulsory involvement of external actors in urban development processes is mostly carried out by the municipality when it serves a specific purpose. It is a selective and stakeholder-targeting participation approach.

It is a common trait that these changes in organization, processes and networks have been intended to be flexible and to make way for entrepreneurial actors—both in the local public and private sector, and in particular, between the sectors. It has been especially through such public–private networks that the significant increase in experience-oriented projects has emerged. This has also implied the emergence of new roles and practices at the actor level. In particular, the municipality has changed its role. The public servants have been encouraged by their political and administrative managers to act as "pioneers" for the remainder of the citizens.[3] Instead of traditional top-down political procedures, more appreciative and action-based methods were prompted and implemented in the municipality–business–citizen relation. The aim was to change the collective mindset of local policymakers and planners from an industrial and bureaucratic approach to an entrepreneurial approach where "everything was possible"[3] so that new urban development projects could emerge. To the citizens and the business community this has mostly been interpreted as a new "great openness and will to find alternative solutions".[2] Increasingly, the municipality is now seen as a project partner instead of a regulatory authority[2] that "plays an intermediary, organising and coordinating role".[5]

> We don't start by finding out what cannot be done. We start by identifying all the things that can easily be done, and then at the end we are left with the real problem. But, then 95% has already been removed. That is a much more positive approach.[3]

The majority of the municipal council "politicians have been willing to take risks—maybe also because they were pushed to the edge by the ship yard closure".[5]

We are more consultants than controllers. We try everything, and we are experimenting. . . . we are not afraid to try things, and we know that not everything succeeds.[4]

Action, Not Planning, as the "Driver" in Urban Development and Planning

The changes described have significantly altered urban development and planning in Frederikshavn. Until the crisis around 1999, urban development in Frederikshavn was mainly influenced by two rather separate spheres; the welfare distributing municipal government and its administration, and a business sector that struggled to maintain momentum in a dwindling local industrial economy. The governing and planning of urban development were primarily matters of municipal land use regulation, and business development was narrow and subject to control. Ideas and projects for change were only allowed to surface and gather political momentum from a few predefined actors. Alternatives to the industrial image and business profile were even tabooed. The crisis helped remove the gridlock and the traditional tensions between the local government, business and to some extent, the citizens.

Today, urban development and planning in Frederikshavn have become the subjects of an extensive urban governance network, where "you cannot point out who decides",[3] because "there are many and they change from project to project".[3] This new broad system of arrangements and coalitions influences a range of public and private investments in urban development projects and business. It interconnects and mobilizes knowledge and resources, and it also influences the formation of new visions and strategies. It is characterized by being flexible, interactive and collaborative. It embraces public as well as private entrepreneurs and experiments, and it is willing to take risks that were earlier "unthinkable".[3] It governs in an indirect and mutually consulting manner—primarily because the municipality politicians and administration act as a transformation agent that "creates space where we trust each other",[1] and because "local business was willing to accept the municipality as an actor in local business development policies".[4]

The urban governance network in Frederikshavn works on the basis of a renewed "Klondike-like" mental framework and through "a set of patterns that we do not necessarily know beforehand". In particular, it is based on the idea that "actions create opportunities, which again creates new actions, etc. That is part of the drive".[1] Furthermore, "the strategy emerges along the way",[5] so that it is an outcome of actions (projects) rather than a starting point. Thereby, the driver has become action, not planning. Many projects (public building investments, festivals, events, palm beach, etc.) have helped to shape a new local identity and development, and lately (2006–2007) this has inspired the emergence of a more coherent strategy for Frederikshavn. The strategy is currently known as "the transformation strategy",[5,1] which transforms the city from its industrial past to a host city that provides events and experiences that both connect the city to its past but also to modern life and the world.

The art is in moving. It is movement that creates development. Creating movement is a goal in itself. . . . We want the change to happen in cooperation with local powers, with respect for history, with willpower and energy, with respect to the spirit of the times and the global development. In this field of tension, the innovation is created that is driven by those who live here. We call that movement for transformation.[1]

Critical Aspects

Local critics have asserted that incipient signs of undesirable emptiness in parts of the city centre are the result of the reduced attention to comprehensive land use planning and regulation. For instance, the former town hall is empty, and a nearby former power plant is also awaiting alternative use. Here, the municipal council wanted to avoid detailed planning and leave room for manoeuvre for potential entrepreneurs. The critics have argued that plans could also be used to attract new interests and investors.[6,7] In addition, the open-ended project approach has also been termed a "spotlight planning" that "lacks logic and coherency" because it "focuses on a very little area".[7] However, recently (2007) the municipal council decided to gather most of the municipal administration on these centrally located areas in a new building complex, in order "to foster synergy ... and cooperation in interdisciplinary environments". The new facilities are also meant to embrace "trade, businesses and other activities that ensures progress and dynamics". Furthermore, this process includes the development of new "holistic plans" for the city centre (Frederikshavn kommune, 2008, p. 3). As such, the municipal council seems to have reinvigorated planning when it needed to create coherence.

In addition, projects are often rushed through to service entrepreneurs, which in some instances causes discontent in the population over legitimacy. This especially occurs when such processes also turn out to be very selective in their participatory approach by favouring only a few stakeholders.[7] It has also been criticized that the maintenance of municipal schools has suffered because of the public investments in experience-oriented facilities.[6] However, in the municipal budget for 2008, schools have now been given a principal political and investment priority. (Frederikshavn kommune, 2008).

Concluding Remarks

Triggered in 1999 by a local crisis with a substantial number of job losses, Frederikshavn entered a process that reinvented its "mental frame" and transformed not only its economic development, but also its identity, image and governance towards an experience economic and entrepreneurial profile.

Municipal (public) investments in urban renewal and later in experience-oriented building projects and activities played a significant part in this process. It illustrates a significant shift in the political focus of the local state; traditional planning elements concerning culture, sports and the urban physical environment were transformed from being mostly part of a welfare providing agenda to becoming increasingly part of an urban growth and business development strategy. Local leisure, entertainment and cultural activities are not only intended for the wellbeing of the citizens of Frederikshavn but increasingly also for adding value, in the shape of experiences, to its guests, visitors, tourists and potential future citizens. Experience-oriented activities and projects offered a credible opportunity for a reoriented economic development as well as the development of a more modern local identity.

The breaking down of old structures and habits also played a significant role. It reflects a shift from a regulatory trouble shooting culture to a more risk-taking, experimental and entrepreneurial culture. In particular, the project-oriented, collaborative and experimenting (verging on anarchistic) approaches offered fertile ground for the production of experiences, because such projects required specific local knowledge and resources from various sectors and actors that were earlier much more separated. The shift in approaches was helped by: (1) internal changes in the municipal administration towards smaller and more flexible units, based on the idea "that slightly too small units force through increased cooperation and coordination between units"; and (2) new networks between the municipality, business, culture, education and selected citizen groups. The municipality has become recognized as a leading transformative agent—a project partner that "starts by saying yes" because this encourages entrepreneurs. Thereby, planning has been turned upside-down—urban strategies and plans have become more directly dependant and an outcome of already initiated or emerging projects. The planning challenge is then to attempt to tie together the projects in order to create coherency.

The Frederikshavn case shows that the changes in old structures and habits were also strongly linked to what we term a renewed "mental frame" that captured and reinvigorated the local culture and identity. A generation's old "Klondike mentality" was rediscovered and given a new face through phrases such as "actions create new opportunities" and "strategies emerge along the way". The rebuilding of strategies emerged from this new mental frame and the action that came along with it. Moreover, action became a tool, not just to urgently initiate new urban development and strategy formation, but also to fill in the emerging new mental frame, e.g. by providing substance to a renewed identity. As such, the case suggests that a mental frame contains ideas of both "who we are" and "how we do things". It contributes with both intellectual and practical cement to keep together the transformation strategy in its ever-changing mode.

However, as the local crisis seems to have played a decisive part in triggering changes, it becomes relevant to consider the continued flexibility and transformative power of the transformative strategy of Frederikshavn. Is the strategy becoming "locked-in" (Healey, 2007, p. 191) so that it fixes so firmly that adjustments to new situations cannot been seen or adapted too? According to the head of the Technical Administration, "by now the crisis is not big enough"[3]:

> The greatest municipal challenge is to create a crisis that is of sufficient magnitude to enable us to open all the mental doors that are necessary in order for us to be at the forefront. . . .We have to be conscious about creating the crisis mood that opens up our mental gateways again. . . . so that we maintain momentum in our identity transformation.[3]

Currently (in 2008), it seems that there is an increased political tension in Frederikshavn between the (new) urban and business growth strategy and the provision of (old) welfare services. For instance, municipal schools have been pushed to the top of the local political (municipal) agenda. However, meanwhile the municipality invests in new business-friendly facilities in the city centre. Tensions have also emerged over the fact that the sometimes a hurried entrepreneurial approach provokes a feeling of being left aside in groups of the public. The extra-compulsory and selective participatory activities are not always seen as democracy-expanding activities, but rather as ways to legitimize swift municipal council decisions. In addition, there are tensions concerned with the trust in the ability of local politicians and officials to create coherency in planning. Such tensions put the transformative strategy to a test. They seem to exemplify a test of the reaches or limits to more risk-taking government-supported neoliberal approaches in urban development and governance—and thereby also to the role of the local state, e.g. could schools be argued as part of an urban and business growth strategy? or could a return to more detailed municipal spatial planning be an instrument for both welfare distribution and attraction of new businesses and citizens?

Notes

1. Andersen, L.R., Head of Culture, Frederikshavn Municipality. Presentation at Roskilde University, 13th September 2007.
2. Christiansen, K., CEO of "Frederikshavn Maritime Erhvervspark" (Business Park), Chairman of Frederikshavn Business Council. Interview, June 2007.
3. Jentsch, M., Head of Technical Administration and Chief Planning Officer, Municipality of Frederikshavn. Interview, June 2007.
4. Sørensen, E., Mayor of Frederikshavn, Socialdemocrat. Interview, June 2007.
5. Sørensen, C., CEO of the Communications Company "Tankegang A/S", Frederikshavn. Interview, June 2007.
6. Christensen, O.R., High School Teacher, Participant in Planning Debates in Frederikshavn. Interview, June 2007.
7. Petersen, J., Pensioner, Participant in Planning Debates in Frederikshavn. Interview, June 2007.

References

Aschér, F. (2002) Urban homogenisation and diversification in Western Europe, in: R. Hambleton, H. V. Savitch & M. Stewart (Eds) *Globalisation and Local Democracy*, pp. 52–66 (Houndmills, Basingstoke: Palgrave).

Bogason, P., Kensen, S. & Miller, H. T. (2004) *Tampering with Tradition. The Unrealized Authority of Democratic Agency* (Lanham, MD: Lexington Books).

Brenner, N. (1999) Globalisation as reterritorialisation: The re-scaling of urban governance in the European Union, *Urban Studies*, 36(3), pp. 431–452.

Brenner, N. (2004) Urban governance and the production of new state spaces in Western Europe, 1960–2000, *Review of International Political Economy*, 11(3), pp. 447–488.

Clark, T. N. (Ed.) (2004) *The City as an Entertainment Machine* (Oxford: Elsevier).

Dryzek, J. S. (2000) *Deliberative Democracy and Beyond: Liberals, Critics, Contestations* (Oxford: Oxford University Press).

Elwood, S. (2004) Partnerships and participation: Reconfiguring urban governance in different state contexts, *Urban Geography*, 25(8), pp. 755–770.

Florida, R. (2002) *The Rise of the Creative Class* (New York: Basic Books).

Flyvbjerg, B. (1998) *Rationality and Power: Democracy in Practice* (Chicago, IL: The University of Chicago Press).

Frederikshavn kommune (1989) *Kommuneplan Frederikshavn Kommune 1989–2000*. Frederikshavn, Denmark: Frederikshavn kommune.

Frederikshavn kommune (1999) *Frederikshavn-området mod 2005*. Frederikshavn, Denmark: Frederikshavn kommune.

Frederikshavn kommune (2001) *Kommuneplan Frederikshavn Kommune 2001–2012*. Frederikshavn, Denmark: Frederikshavn kommune.

Frederikshavn kommune (2008) *Informationsudgave—budget 2008*. Frederikshavn, Denmark: Frederikshavn kommune.

Hajer, M. A. (2005) Setting the stage: A dramaturgy of policy deliberation, *Administration and Society*, 36(6), pp. 624–647.

Hajer, M. A. & Versteeg, W. (2005) Performing governance through networks, *European Political Science*, 4(3), pp. 340–347.

Hajer, M. A. & Wagenaar, H. (Eds) (2003) *Deliberative Policy Analysis: Understanding Governance in the Network Society* (Cambridge: Cambridge University Press).

Hall, T. (1998) *Urban Geography* (London: Routledge).

Hannigan, J. (1998) *Fantasy City* (London: Routledge).

Hansen, C. J. (2006) Urban transport, the environment and deliberative governance: The role of interdependence and trust, *Journal of Environmental Policy and Planning*, 8(2), pp. 159–179.

Healey, P. (2007) *Urban Complexity and Spatial Strategies: Towards a Relational Planning for Our Times* (London: Routledge).

Jessop, B. (2000) *Globalisering og Interaktiv Styring* (Frederiksberg: Roskilde Universitetsforlag).

Kooiman, J. (Ed.) (1993) *Modern Governance: New Government–Society Interaction* (London: Sage).

Lorentzen, A. (2007) *Frederikshavn Indtager "The Global Catwalk"*, Working Paper, Aalborg: Department of Development and Planning, Aalborg University.

Lorentzen, A. (2008) *Knowledge Networks in the Experience Economy*, Working Paper Series No. 319, Aalborg: Department of Development and Planning, Aalborg University.

Lorentzen, A. (2009) Cities in the experience economy, *European Planning Studies*, 17(6), pp. 829–845.

Marling, G. & Zerlang, M. (Eds) (2007) *Fun City* (Copenhagen: The Danish Architectural Press).

Metz, T. (2007) Fun! Leisure and landscape, in: G. Marling & M. Zerlang (Eds) *Fun City*, pp. 23–34 (Copenhagen: The Danish Architectural Press).

O'Dell, T. & Billing, P. (Eds) (2005) *Experiencescapes* (Copenhagen: Copenhagen Business School Press).

Pine, J. B., II & Gilmore, J. H. (1999) *The Experience Economy* (Boston, MA: Harvard Business School Press).

Pløger, J. (2004) Planlægning i en kompleks og plural verden—og for den meningsfulde by, in: H. S. Andersen & H. T. Andersen (Eds) *Den mangfoldige by—opløsning, oplevelse, opsplitning*, pp. 169–185 (København: Statens Byggeforskningsinstitut (SBi)).

Romein, A. (2005) The contribution of leisure and entertainment to the evolving polycentric urban network on regional scale: Towards a new research agenda. Paper for the 45th Congress of the European Science Association, Amsterdam, 23–27 August.

Schulze, G. (1992) *Die Erlebnisgesellschaft: Kultursoziologie der Gegenwart* (Frankfurt: Campus Verlag).

Sehested, K. (2003) *Bypolitik—mellem hierarki og netværk* (København: Akademisk Forlag).

Smidt-Jensen, S., Skytt, C. B. & Winther, L. (2009) The geography of the experience economy in Denmark: Employment change and location dynamics in attendance-based experience industries, *European Planning Studies*, 17(6), pp. 847–862.

Therkildsen, H. P. (2007) *Nytænkning i byplanlægningen—Frederikshavns forvandling*, Working Paper Series No. 320, Aalborg: Department of Development and Planning, Aalborg University.

Toffler, A. (1970) *Future Shock* (New York: Random House).

Index

Page numbers in *Italics* represent tables.
Page numbers in **Bold** represent figures.